The Enlightened Gene

THE ENLIGHTENED GENE

Biology,

Buddhism,

and the

Convergence

That Explains

the World

ARRI EISEN

YUNGDRUNG KONCHOK

FOREWORD BY HIS HOLINESS THE DALAI LAMA

ForeEdge

ForeEdge
An imprint of University Press of New England
www.upne.com
© 2018 Arri Eisen and Yungdrung Konchok
Manufactured in the United States of America
Designed by Eric M. Brooks
Typeset in Whitman and The Sans by Passumpsic Publishing

For permission to reproduce any of the material in this book,
contact Permissions, University Press of New England,
One Court Street, Suite 250, Lebanon NH 03766;
or visit www.upne.com

Library of Congress Cataloging-in-Publication Data
available upon request
Hardcover ISBN: 978-1-5126-0000-1
Ebook ISBN: 978-1-5126-0125-1

5 4 3 2 1

To

GABE, MICAH, LISA

and

NAMKHA GYALPO AND LHAMO BAIJI

CONTENTS

THE DALAI LAMA

FOREWORD

This book, *The Enlightened Gene: Biology, Buddhism, and the Convergence That Explains the World*, represents something close to my heart. Even as a young child in Tibet I was curious about how things work. But at the time I had no access to formal scientific instruction. However, my formal Buddhist education, based as it was on the ancient Nalanda tradition, emphasized critical analysis and the employment of logic and reasoning.

Living in exile brought me into contact with many people from many different walks of life, scientists among them, which rekindled my interest in how science explains the world. Eventually, regular discussions between scientists, particularly experts in cosmology, neurobiology, physics, and psychology led to my establishing the Mind & Life Institute in collaboration with Francisco Varela and others.

As discussions progressed, it became evident that monastics needed more specific training to be able to participate in dialogue with scientists. Monastic study of modern science began with courses and workshops. In response to an evident need, Emory University began to collaborate with the Library of Tibetan Works and Archives on developing and implementing a systematic plan for educating monks and nuns in modern science. Textbooks and video lectures have been created and translated into Tibetan. A new scientific terminology has been developed in Tibetan, and monastic science teachers have been trained. The result is that a refined science curriculum has now been fully implemented in Tibetan monastic centers of learning.

This book, *The Enlightened Gene*, represents real collaboration between a modern scientist and a Tibetan monk-scholar. Dr. Arri Eisen

ix

has been involved with developing the curriculum of the Emory-Tibet Science Initiative from the start. Yungdrung Konchok, a monk of Menri Monastery, a graduate of that initiative, and a Tenzin Gyatso Science Scholar, spent three years studying science at Emory University. Their work represents not only scholarly achievement but also the friendship and understanding that can come from open dialogue between great intellectual traditions.

I would like to congratulate the authors on their achievement. I am confident that this kind of collaborative work will add to the advancement of human knowledge, leading in the long term to a more holistic understanding of what it is to be human. I have no doubt that all who read this book will benefit from the insights generated by the convergence of science's understanding of the material world and contemplative traditions' understanding of the workings of the mind.

The Enlightened Gene

Prologue

When the student is ready, the teacher will appear.

BUDDHA

 n just my short twenty-five-year career, biology has dramatically changed. To keep up with the breadth, depth, and amount of new knowledge generated these days—even one small corner of it—is nearly impossible. But imagine entering the scene from an entirely different world, one in which you have barely a scrap of previous exposure to science in your six-hundred-year tradition of learning. How would the nature of your learning change? The nature of the teaching? How would the science itself change?

This book is about biology and Buddhism. It's about how an unusual project involving American scientists and Buddhist monks can enlighten us in teaching and learning across worldviews and in general. Two leaders of the project—one a scientist and one a monk—tell the story.

DHARAMSALA: 2011

The Dalai Lama sat before us on a big wooden chair. This was the annual audience to update him on our project teaching science to Tibetan Buddhist monks and nuns. In 2005 the Dalai Lama invited Emory University in Atlanta to develop and establish a modern science program to become part of the centuries-old curriculum of his twenty thousand monastics in exile. Through his lifelong interest in science and his recent conversations with neuroscientists, he saw the great potential for alleviating suffering and enriching humanity by integrating cutting-edge science with ancient wisdom, while at the same time engaging monks and nuns in twenty-first-century knowledge.

We were at the Dalai Lama's home in Dharamsala, India, where he has lived since escaping from China over the Himalayas in 1959. Prior to his escape, he was isolated from the world in a Tibetan palace from the time as a young boy when he was identified as the fourteenth reincarnation of the Dalai Lama. Now, still considered a god by many of his own people, the Dalai Lama is a Nobel Laureate recognized the world over as an icon of peace and compassion. At our meeting, I was to present him with the science textbooks we had written thus far, books translated into Tibetan to catalyze our project as it moved into the monastic universities.

When, six years before, my friend Geshe Lobsang Negi asked me if I would help respond to the Dalai Lama's request to teach modern science to the Tibetan Buddhist monastics in exile, I jumped at the chance. I had been teaching biology, ethics, and science and religion for nearly two decades, and here was something clearly new and different. Little did I know how this project would change me.

I grew up Jewish in the Baptist South of the 1970s. Both my parents are teachers—my father in science, my mother in English. My dad has been a scientist for more than forty years. When I was a little kid, I would go with him to the lab. He was studying the genetics of obesity in mice.

There we were: two very skinny people studying fat mice. I am sure my father's interest in science encouraged me to become a scientist myself. I remember doing an experiment with him to see how well mice lived on breakfast cereal versus regular mouse food. The answer: not very well.

"And now, Your Holiness, Dr. Arri Eisen will present to you some of the texts he has written, which we have translated into Tibetan for the monks and nuns, as our project continues . . ."

I snapped out of my reverie.

The Dalai Lama accepted the books, wrapped in the white ceremonial scarf, or *kata*, from me and held them as he reiterated one of his core messages: it is education that we need, it is education that will change the world—not meditation, not religion, but education.

YUNGDRUNG KONCHOK has been involved in the project since nearly the beginning. Konchok grew up three days' walk from anywhere in Nepal and entered a monastery when he was fourteen; he was twenty-five when we first met. Konchok is now in charge of his monastery's extensive library. He is one of several monks who, in addition to working with us in India in the summers, also studied for three years at Emory University, taking undergraduate science courses. Konchok and I have become especially good friends.

Konchok took two of my classes—introductory biology and cell biology—at Emory, and every Friday morning for more than two academic years, Konchok and I met at Starbucks to catch up and share stories. To passersby, we must have looked a pair: a tall, angular, white professor-type hunched over hot drinks with a small, bespectacled, maroon-robed monk in an Atlanta university coffee shop. This book was born during those meetings, developed at Emory and matured by e-mail and more Friday meetings—this time via Skype—between Atlanta and Konchok's monastery in the Himalayas.

SEATTLE: 1986

How did I wind up as a teacher, much less one teaching science to Tibetan Buddhist monks and nuns?

Looking back, I see that the seed for both was planted a quarter century earlier:

Outside the sixth-floor window, it was midnight, and the bright lights of the massive Interstate-5 bridge sparkled across the water as it flowed off toward Lake Union and the future. The lab was quiet, and I was alone with my tubes and pipettes, tending carefully to paper membranes coated with DNA or RNA, the codes of life.

I was in my first year of graduate school—the PhD program at the University of Washington in Seattle—twenty-two years old and gung ho, learning the secrets of molecular biology in a lab in the thick of it at that moment.

Suddenly, I was blindsided by an overwhelming wave of emotion: I had no idea what I was doing. I knew the steps of the experiment and I knew how to do it, how to follow the recipe. But I did not know why

I was doing what I was doing. Why did this experiment matter? And not just in an existential way, but in a biological way. What was its substance, its context? How did this experiment fit in with all the experiments that came before it, and what would it mean if the result tells us this versus that? Exactly what big questions did my little question about a few genes in a mouse address? How did these genes relate to the rest of the genes in the mouse? To genes in humans? To biology in general?

I sat down, a bit stunned.

In the following months and years, that night kept returning to me, bothering me more each time. It disturbed me because I had done everything one was supposed to do in the United States to become a card-carrying scientist. I had been lucky enough to attend excellent schools and earn good grades and get into one of the best graduate programs.

If someone like me, a "successful" product of the system, was so profoundly lost, what did this say about the system? Did I shine in the system, did I survive, only because of a photographic memory and a natural attraction for science? If I didn't really know what my small experiment *meant*, much less what science is in the greater context of history, ethics, and society, I had a problem. And I began to think science and science education as a whole had a problem.

Although it did not consciously register at the time, that night in the lab was the moment when the path toward becoming a teacher opened before me. I would ensure that others did not wind up like me—"successful," but not clued in.

ATLANTA: 2005

The Dalai Lama is a professor at Emory, and we have a strong program in Tibetan Studies, which features a semester-long program for undergraduates in Dharamsala.

The Dalai Lama requested that our new science curriculum for the monks and nuns focus on the life sciences (with a special emphasis in neuroscience) and physics. About a dozen professors from these areas and from Tibetan Studies began meeting at Emory. We started from scratch: How do we teach science to educated young men and women who know how to learn, but who know very little science or math?

How much knowledge in science did the monks and nuns actually have? What aspects of Tibetan culture and history and Tibetan monastic culture were important to know?

The Library of Tibetan Works and Archives in Dharamsala, directed by Geshe Lhakdor, a former translator for the Dalai Lama, became our partner in the Tibetan community. The library's main charge is to preserve the culture of Tibetans in exile, conserving ancient texts and translating new texts into Tibetan, developing websites, and hosting classes and workshops. Traditional Tibetan texts, often smuggled over the Himalayas in sacks, are not bound and vertical like books in the West; instead, they are stacked pages of handwritten parchment, unbound, resting horizontal, and wrapped in fabric on the library's gray metal shelves.

Our project working group met regularly for two years. We consulted with Americans who had become Buddhist monks and with other scientists who had taught monks. We talked to translators and thinkers and writers in this area—including B. Alan Wallace, who is an American and was a Buddhist monk and who has been involved in many science and Buddhism discussions, and Thupten Jinpa, an accomplished Tibetan scholar and translator for the Dalai Lama. We met with representatives from other projects, notably Science for Monks and Science Meets Dharma, who years before had begun work with Tibetan monks and science.

We began with a five-year pilot curriculum for monks and nuns from diverse monastic institutions who were especially interested in science. During this time, we tested ideas, developed and translated texts, and got acquainted with each other's cultures and learning styles. Following this pilot, the program would, beginning in 2014, move into the three major monastic universities in southern India, where the majority of Tibetan Buddhist monastics in exile live and study.

Two different cohorts of thirty-five and fifty-five monks and nuns, respectively, completed the pilot curriculum. Each summer for several weeks we all met in Dharamsala. Teams of scientists taught each area —life sciences, neurobiology, or physics—for one to two weeks, one after the other with breaks in between. During the classes, every few sentences the teachers stopped, and our words and ideas were translated into Tibetan. Classes met for six hours a day.

Often Tibetan Buddhist monks and Americans have the same questions.

If you ask the average American about microevolution—say, could Darwin's finches' beaks evolve into different shapes to eat different kinds of seeds on the Galapagos Islands?—they say, sure. But getting life from no life, the origin of the first cell? Or claiming that humans and chimps share a common ancestor?

In retrospect, I see why I was so attracted to this Emory-Tibet project. The Dalai Lama and Konchok are interested in how science and medicine can help people of faith, while I am interested in how faith and belief can help people of science.

In the United States, there is an idea that religion and belief (and for some, even ethics) should be excluded from science—actually excluded from the academy altogether. But this works to the detriment of scientists and science in a country where religious belief is very strong. I teach science to some of the best undergraduates in the United States, the nation's future physicians and researchers. When I ask them, say, if they believe in evolution, nearly all of them say yes; then when I ask the same students if they believe something in addition to evolution had a role in making the human species what it is, half or more of them say yes. So why ignore the elephant in the room? Why should scientists not address the beliefs of our students head-on?

Maybe teaching monks science in Dharamsala and undergraduates science in America is not *that* different. In America, we often teach across different cultures, religions, assumptions, knowledge, and experiences. In my lab are men and women who grew up in Egypt, China, India, Italy, and the United States. My students, whether I am teaching science or ethics, undergraduates or physicians, are at least as diverse.

The original hope of the Dalai Lama was that our project would integrate ideas and practices of modern science and Tibetan Buddhism —especially modern neuroscience with ancient mind-body knowledge —to help relieve suffering in the world. At the same time, he realized monks and nuns must understand the basics of science to be effective citizens of the twenty-first century. We have made strides in these directions, as well as into many unexpected areas, including learning across cultures and within different belief systems within those cultures. We are learning how approaching the same information with a shift in perspective can dramatically change how we explore and expand that information.

Here is an irony: American scientists are often less open to the possibility that ideas from Buddhism or other religions have relevance to their work than the Dalai Lama and the monks and nuns involved in our project are open to Western science and medicine having an impact on their beliefs. And scientists are the ones who often accuse people of faith of being closed-minded and not listening to reason.

Perhaps one of the legacies of the Emory-Tibet Initiative, as it expands into the monasteries and the sciences become a permanent part of the monastic curriculum, will be to provide a model for effective interaction and occasional integration between science and faith communities.

Perhaps moving the discussion out of the Judeo-Christian forum, out of what we Westerners know best, will make the conversation easier. Then we can go back and apply lessons learned.

THE ENLIGHTENED GENE uses our unique project as a lens into a cultural moment, joining Americans' longtime fascination with "the wisdom of the East" and very recent quantum leaps in biological knowledge. Many of these new biological discoveries can actually be read as moving science *closer to spiritual concepts*, rather than farther away. We use our personal experiences and knowledge, our stories, in the hope of opening up and laying a foundation for serious conversations, integrating science and spirit in tackling life's big questions.

While Americans trust and support science and medicine, there is clearly a sense that something is missing in their practice and application. Much of twenty-first-century American political discourse, on such issues as climate change, abortion, genetically modified organisms, and evolution, is often framed in a science *versus* spirituality and religion context. It's as if you have to choose between science and religion, one or the other. We turn such thinking on its head. A recurring theme of the book is that science often "discovers" significant truths that Buddhism and other religions have long known (but, importantly, from a different angle and without the measurability of science).

The backdrop of teaching science to Tibetan Buddhist monks and nuns allows exploration of how science and religion—how dramatically different worldviews in general—can work together to enrich each other, to shed light on life and what it means to be a thinking, biological human being.

Our book emerges from years of experiences and interactions and hours and hours of one-on-one conversations between Konchok and me. The bulk of the resulting text is written by me, but is based on our discussions and shared writing and is enhanced with direct insights and reflections from Konchok sprinkled throughout. Konchok is also an accomplished artist, and our book features his illustrations.

Each chapter of our book integrates Buddhism and biology and uses striking examples of how doing so changes our understanding of life and how we lead it.

CHAPTER 1, "Are Bacteria Sentient?," demonstrates how a Buddhist perspective on bacteria—organisms we usually think of as little single-celled beasts to be eliminated—helps us understand how they are actually more often helpful, even vital, organisms, necessary for our thinking, eating, and otherwise living life as we know it.

CHAPTER 2, "Life, Death, and Sacrifice," illustrates the essential nature of death within life—the Buddhist worldview of cycling, self-sacrifice, and altruism—at every step of biological development, from sperm and egg to adult, from the molecular level to the very nature of our brains and how we learn.

CHAPTER 3, "How Did Life Begin?," tackles evolution, the most contentious of science and religion issues in the United States, and uses the question of life's origins to explore both biology and how our project is a model for science and religion productively engaging, rather than gutting, each other.

CHAPTER 4, "Altitude and Attitude," elaborates the startling physical and mental adaptations of Tibetans as a lead-in to unveiling novel, extraordinary research on how genes and environment together shape who we are and what we become. We explore epigenetics and mix it into the story of Tibetans' and all humans' remarkable resilience.

CHAPTER 5, "Ecology and Karma," shows the amazing extent to which new ways of thinking about ecosystems resonate with the central Buddhist concepts of karma and interdependence, catalyzing new insight into conserving our planet.

CHAPTER 6, "Are Humans Inherently Good?," provokes a discussion of the inherent nature of human goodness, to link Buddhist thought and recent research exploring the effects of meditation on empathy and compassion.

CHAPTER 7, "Meditation and the 'New' Diseases," investigates, through research performed in the context of our project, how ancient Buddhist practice shows promise in both relieving and preventing stress and stress-related diseases, that is, the diseases Westerners now most commonly suffer and die from.

CHAPTER 8, "Beyond Science and Religion," uses a conversation with the Dalai Lama to look back through and synthesize the key concepts in the book as a way to look forward to what's next in globalized twenty-first-century science and religion.

DHARAMSALA: 2007

To teach my biology class today, I traveled almost halfway around the world on three planes and in three cars.[1] Outside the classroom door, dozens of sandals rest in neat rows. I take off my tired New Balance shoes and walk in.

My luggage has still not arrived. I have discovered with the help of a number of tickled Tibetans that in Dharamsala it is impossible to find underwear (or any clothes for that matter) for someone like myself, well over six feet tall.

Forty Tibetan monks and nuns sit cross-legged on the floor (although it is very unusual for monks and nuns to study together), their shaved heads emerging from maroon robes. In the years to come, I would get to know them well, their individual strengths and eccentricities. I couldn't have imagined then how my life would interweave with those in that room—how Dhondup would provide insights to change my teaching, Ngawang would dramatically alter relationships among Chinese and Tibetans back in Atlanta, Sangey would live with my family back in America and become a lifelong friend, and Konchok and I would write this book. But at this meeting, I am barely able to differentiate them; it is even difficult to tell monks from nuns.

With my body clock nine time zones away, I had awoken a few hours earlier at 5 AM in a modest dorm room, a sturdy ceiling fan whirring away. When I poked my head outside the door, a lone monk was pacing the flat roof across the way, five floors up, saying his prayers, embraced by the rays of the sun climbing over the Himalayas.

We are in a small college near Dharamsala, India, nestled into the foothills of these majestic mountains, the home of the Dalai Lama and the Tibetan government in exile for half a century. The classroom is cinderblock spare. The monastics sit, notebooks perched on their knees. In the afternoon, the sun bakes us from behind the blinds, easily canceling the efforts of the two small air conditioners and half a dozen ceiling fans. There are a whiteboard, a computer, and a projector. Electrical power is intermittent.

I introduce myself through our translator, Sangey. I tell them about myself, about America, my family, and my adventures on the way to Dharamsala the day before. I had barely fit into the "van" that came to pick me up from the airport, and we were lost for two hours in Delhi in the middle of the night looking for my hotel. I was alone, severely jet-lagged, and unprepared for the chaos of streets packed with people at 2 AM, cows aimlessly roaming the highways, stoplights ignored, endless streams of animal traffic, and the barefoot man pushing a cart stacked ten feet high with plastic containers.

The Buddha says there are two mistakes you can make along the road to truth—not going all the way and not starting.

Class begins.

CHAPTER ONE

Are Bacteria Sentient?

At first I thought cells were just rudimentary pieces,
like bricks; now I realize each cell is a universe.
Without cells there would be nothing.
KONCHOK

acteria of all sorts, some appearing as big as grapes, dance and spin on the pale cinderblock walls of the Dharamsala classroom. The monks and nuns stop and point and exclaim. Something has changed; the room shifts.

We forget about the afternoon heat pressing down on us and stare excitedly at the bacteria projected in real time from the microscope slide onto the wall before us.

Was this what it was like for the original microbe hunters—Leeuwenhoek, Spallanzani, Hooke, and their ilk—when they first uncovered this astonishing world centuries ago? Imagine the thrill and the fear of discovering that so many zillions of creatures existed everywhere, all around us, all the time. It must have been almost beyond belief.

These particular Dharamsala bacteria, these single-celled "beasties" as Leeuwenhoek called them, were grown by the monks and nuns in the class (some of the drawings of the bacteria Konchok made that day are in figure 1.1). They swabbed the beasties from doorknobs or fingernails and nurtured them on bacterial food plates whipped up from cornstarch we found in the campus kitchen.

One of those microbe hunters, Robert Hooke, coined the word "cell" in 1665; when he saw those tiny walled spaces in the tissue of cork that he was the first to ever see, they reminded him of the cells monks live in.

FIGURE 1.1 Konchok's sketches of the first bacteria he ever saw, projected from the microscope onto the classroom wall in Dharamsala.

In our group discussing whether bacteria are sentient beings, we had two opinions; we were split. Before the experiment, I myself was thinking, "Bacteria are not sentient."

But when we saw the images and the bacteria moving toward food on the wall, then I thought this was real. I thought they could be sentient. Before this, I thought just because they move and find food, this doesn't mean they have to be sentient. But in the microscope, it looked like the cells had a purpose. After our experiments, we monks talked a lot. I thought: Maybe these bacteria have senses, maybe even emotions. They "feel" where food is. Maybe they *are* sentient.

This was the first lab experiment we ever did; we saw the bacteria on the wall.

Most of us got it then—what scientists do. In actually doing it, not just saying it, many monks changed their minds.

Back in my monastery, the monks always ask, "How do you do experiments? How do you use the equipment?" In our own philosophy, we debate on what Buddha taught, we explore the logic of texts and rationality.

When you say scientists hypothesize, experiment, analyze, monks say "how?"

When we did this experiment, things made sense. This became strong evidence. We had some negative ideas taught to us about science, but that experiment clarified some doubts about science.

Some monks didn't believe in science. But when we did this experiment, we saw this with our own eyes, a kind of truth. Not only that, the evidence really inspired us a lot. That experiment motivated us to learn more science and explore more.

Our goal is to create a lens, literally and metaphorically, through which information is experienced in a rich context. The monks and nuns learn about cells and the cell theory—that cells are the fundamental unit of all life—and they learn this in the larger context of the themes of biology and of how scientists ask questions and approach problems.

Why should Tibetan monastics care about cells? How is it related to their lives and experience? This is where the question of bacterial sentience comes in; it engages a core Buddhist concept. Buddhism teaches that sentient beings are aware creatures, such as humans and other animals. Compassion should be shown to all sentient beings, and any such being can be reincarnated as any other. So, *if* bacteria are indeed sentient beings: (a) any person might be reincarnated as one, (b) any bacterium might be reincarnated as a human, and (c) we should show compassion to all bacteria. At first, many of us Western scientists might dismiss out of hand as trivial, silly, or irrelevant, the question of whether bacteria are sentient. But as we'll see, such an attitude is, and was historically, perhaps to our ultimate detriment.

The monks and nuns, then, have a vested interest in exploring the question of whether bacteria are indeed sentient. And we have established a culturally relevant lens through which to teach cell biology. In order to answer the sentience question, other questions, at the heart of cell biology, must be addressed: What are cells and of what are they made? How do cells sense, translate, and respond?

We wrestle and walk with these questions six hours a day for eight days with discussion and laboratory experimentation. Science is about experiments. The bacteria sentience question is the monks and nuns' door into experiments.

BACTERIA ARE a powerful model system. A model system is an organism studied in order to ask fundamental biological questions. All living things are made of cells, and the basic constituents of cells have been conserved throughout evolution. Questions relevant to more complex organisms like humans can often be addressed in relatively simple organisms such as bacteria, flies, or yeast. These creatures also are cheaper to work with, are easier to maintain, have short life cycles, and garner fewer ethical concerns than "higher" organisms more similar to and engaged with humans, such as chimpanzees or dogs.

Experiments on animals such as mammals also, of course, occur in the West. And the question of which model systems to use, or whether

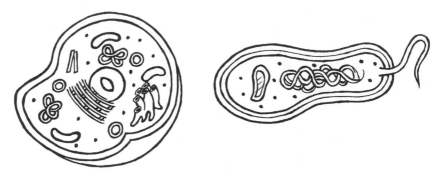

FIGURE 1.2 A typical human cell on the left and a typical bacterial cell on the right.

to use any at all, resonates especially with Buddhists, who show compassion to all sentient beings.

The monks and nuns are not comfortable conducting experiments with anything living—unless it is plants, which they do not consider sentient, or something like bacteria, the sentience of which is the question under investigation. The monastics would not kill any organisms, but do, in our Dharamsala classes, study organisms that are already dead.

So our choice of bacteria as a model system to explore in Dharamsala is effective, because studying bacteria demonstrates the power of experimentation and the importance of model systems without introducing, at least initially, the problem of studying sentient beings.

Those bacterial cells dancing on the wall (and at first seemingly so different from our own cells) are in many ways very similar to human cells. Figure 1.2 shows a typical human cell on the left and a typical bacterial cell on the right (not drawn to scale). And because they are ubiquitous, easy to access, and grow quickly, and their maintenance and study is cheap, bacteria are excellent for asking fundamental questions about cells—even with limited supplies in a makeshift lab in the foothills of the Himalayas.

During my study of Western science, I realized that it promoted my knowledge and wisdom, and developed my insight. Simultaneously, it has given me the desire to learn more and made me more inquisitive in whatever activities I do.

In fact, desire and grasping, craving, wanting more is usually considered something undesirable, something to be given up in Buddhism. But knowledge is an exception to that rule; learning new things is an activity that one should engage in all one's life.

In some circumstances the Buddhist concepts of nonvirtue, ethics, and sin restrained me somewhat in accepting some of the scientific explanations and methods. For instance, in Buddhism you do not kill sentient beings or make them suffer; and as a monk, I took a vow to remain celibate and not to kill, steal, or lie. But in the laboratory, science students are required to design and perform experiments using animal-based model systems. Dissecting these formerly living beings and interacting with them is one of the most effective methods of achieving many scientific goals, that is, examining firsthand their anatomy and learning much about the causes of birth and death. I have found it personally difficult to perform dissections and manipulation using real animal models. Essentially, all animals are our kin, possible past or future versions of our own reincarnation, so making them suffer makes us suffer.

When I took introductory biology at Emory, we used zebra fish in experiments in the laboratory. It was interesting to observe and play with the different stages of zebra fish development, but Buddhists believe that dissecting and using an animal or insect for experiments is nonvirtuous and unethical.

The Buddha encouraged his disciples to have respect for all life forms and not to unnecessarily damage or destroy any living thing.

"Should scientists use animals in research?"

The Dalai Lama answers in what I have come to learn is his classic style. He echoes Buddha's teachings: respect all sentient beings, and do not intentionally harm or destroy them. *But,* the Dalai Lama also says, we must examine each case carefully and balance the issues at play. What if experiments on ten mice could potentially save the lives of thousands of humans?

This question about animal research comes from a student at a teaching of the Dalai Lama's I am attending at a Tibetan Children's Village in Dharamsala. These villages in India adopt and raise Tibetan children who escaped, often without their parents, from their home country and help transition them into society. Thousands of such children escaped in the latter half of the twentieth and early twenty-first centuries, but the route became more dangerous and often fatal as a result of more active Chinese military intervention after the unrest in Tibet in 2008. One of the young Tibetan students asks the question.

I am struggling to follow the translation of the question and the Dalai Lama's answer through earphones plugged into a transistor radio, like the kind I used to sneak into my junior high classes to listen to college basketball tournament games in the 1970s.

I try to focus on the Dalai Lama's response, but my legs went numb an hour previously. I was shown through a side door by the Dalai Lama's sister to sit on the stage where he would talk. This sounded good, until I saw I would be sitting on a floor mat cross-legged for the next three hours—a less-than-exciting prospect for an inflexible six-and-a-half-foot-tall American.

We had wound our way up to the village in a tiny car on one of Dharamsala's unimaginably narrow roads on which cows, buses, trucks, innumerable motorcycles, and pedestrians—locals and truth seekers of every stripe—perform their daily death-defying waltzes (as I move forward toward you, you simultaneously move toward me into the space I just vacated, and vice versa). If you mess up, it's a very long way down the mountain. Also, don't forget, just to add more spice to the adventure, you have to keep track of the fact that Indians drive on the left-hand side of the road.

When the Dalai Lama escaped from Tibet over the Himalayas in 1959, the prime minister of India at the time, Jawaharlal Nehru, offered him—and all the other exiled Tibetans in the years since—different communities and resources to maintain their culture. Dharamsala, where we initiated our project teaching science to monastics, became the home for the Tibetan government in exile. Other lands in southern India, where we now teach in three large monastic universities, were established as refugee communities.

WE CHALLENGE the monks and nuns in our class to explore how they might "ask" the bacterial cells themselves to address the sentience question.

The monastics struggle first with a definition for "sentience." Does it mean merely "the capacity to sense," or does it imply some higher level of awareness? Tibetan Buddhism holds that plants *are not* sentient. Are bacteria different from plants? Might science have one definition for "sentience" and Buddhism another?

We settle on a set of experiments that will at minimum address the question of whether bacteria can *sense*. The monks will grow bacteria and, while observing them under the microscope, will add different chemicals, one per experiment, to one side of the microscope slide to see if the bacteria show any clear change in movement away from or toward the chemical. They will use chemicals that are predicted to attract, such as sugar, as well as some predicted to repel, such as acid.

We make the bacteria-food plates, and the monks and nuns swab different surfaces to pick up bacteria and then transfer them to the plates. In Dharamsala's very warm summer temperatures, the bacteria grow quickly, and soon our makeshift laboratory smells like a neglected locker room.

Western science says each species lives in a certain kind of environment— fit for itself. An ant fits his environment. He needs food; in order to get food, he has to run out, to search for it.

In Buddhism we say all sentient beings are like our parents or brothers. The question of killing in science is different; it's more of an ethical than a religious problem.

Buddhists say that an ant is a life, a sentient being. It is not good to take life. It's very negative. In Buddhism, an ant from many thousands of years ago could have been our relation.

In terms of evolution, you could say all animals are brothers, as ants use many of the same genes and proteins that we have to do similar things. But we also believe animals have the same senses and emotions that we do—fear, hope. All sentient beings seek happiness as we do. And not only do they feel the same, but *they have the same life goals that we do.*

His Holiness the Dalai Lama says that if many thousands of people could be saved by doing tests on a few animals, maybe that would be okay. In Buddhist texts, it says we should not tell a lie, except in certain cases. For example, if a hunter is chasing a deer, and you see the deer run by you, and then the hunter comes and asks if you saw the deer run by, you must point him in the opposite direction than the deer actually went.

Two kinds of translation happen simultaneously in our Dharamsala classroom. We teach the monks and nuns biology. As we teach, translators take our English words and science concepts and turn them into Tibetan words and concepts the monastics, we hope, are able to understand; the monastics in turn ask us questions, which are translated into English and to which we respond.

Analogously, the bacteria in the experiments, under the microscope and projected on the wall, take the information from their environment—chemicals in their medium, other bacteria in the neighborhood—and translate it into particular behaviors.

ALL TEACHING is a kind of translation; good teaching is good translation. To translate effectively, teachers must know their students. Before going to Dharamsala, we spent two years studying Tibetan Buddhist culture and talking with Tibetan monks and educators who had worked with them.

How could we teach science to intelligent people who have not had a significant change in their curriculum—focused on Buddhist philosophy and ancient texts—for six centuries, people from as different a world from ours as imaginable, a world wary of science?

How could we teach science well to people who know absolutely no math, who have never thought in terms of scientific experiments, who have rarely written a paper or been on the Internet?

How could we teach modern science to people who believe in reincarnation?

Often we think of ideas and information being *lost* in translation. The number of formal and informal languages in our project with the monks and nuns—Tibetan, Hindi, English, Nepali, Kanada, Chinese, Mongolian, mathematics, neuroscience, monk, physics—is mind-boggling.

Initially, the translation heaviness of the project seemed an obstacle. Now, nearly a decade later, I see *the opposite* is true. At first, we bemoaned the fact that we would not be able to cover much material with the monks and nuns, because we had to stop every few sentences for Tibetan translation.

The process of translation, as it passes from the teachers' cells and organs to the monks and nuns, to the translators, and back again, creates time and space for reflection and deeper understanding. There is a strong feeling among the monastics that the translators are much more than technicians connecting English to Tibetan and scientists to monastics, but that they are also "restless mediators," listening carefully to the teachers, explaining what they say clearly, carrying the monastics' questions to the scientists, and bringing the answers back.

AS THE BACTERIA isolated by the monks and nuns grow, we consider *how* these cells are growing and responding to their environments on the plate. On the broader level: what makes a cell a cell? We teach the parts of the cell in the context of the sentience question and in the context of the bacteria experiments the monks and nuns design.

The original cell theory, developed by German scientists in the 1800s, states that the cell is the basic unit of life, all living things are composed of cells, and all cells come from previously existing cells.

We discuss with the monks and nuns what other characteristics all cells, and by extension all living things, must have. They see their bacteria growing dramatically overnight. Cells grow and divide; therefore, cells must be able to find and use energy, and they must be able to pass on information to the next generation.

To do all these things, cells must sense and respond to their environment and adapt their behaviors to fit new situations. At the same time, cells need the capability to be able to return to a stable state after an environmental change, say, in temperature, nutrition, or presence of other related or unrelated, helpful or dangerous organisms. The maintenance of stability, called homeostasis, introduces the theme of regulation.

Homeostasis is maintained by feedback regulation that occurs within many biological cycles. Feedback—when a product of a process circles back to affect the process that creates it—is a common and effective regulation strategy.

Cycles are as central to Buddhism, as we will see, as they are to biology. Samsara, the great cycle of life and death, comes to mind, as do the many faithful pilgrims circumambulating the temple near the Dalai Lama's home in Dharamsala.

Cells are really powerful. Without cells there is nothing. Cells are building blocks. They are like the concept of *sunyata*. *Sunyata* is like nothing *yet*, but with a lot of potential. *Sunyata* is like potential energy, nothing yet material, but energy that can become material and concrete. *Sunyata* always existed —before the formation of the universe and now, not creating anything exactly, but being used to create things. We use this concept of *sunyata* in our meditation practice to visualize or think of such empty potential—no attachment, no desire—that can bring you to a space of goodness. When you think of reality, you automatically think of desire and attachment. I think of cells like this: the potential for something new, for innovation. Then the cell is the beginning.

For me, learning about cells changed a lot. You can say a cell is a kind of brain; it retains memory, gets signals, gives signals.

If the cell is somehow aware, how? To develop a story of how cells might accomplish some level of awareness, as their bacteria grow, the monks and nuns work in small groups to draw models on paper. Draw bacterial cells and what you think they might require in order to sense, translate, and respond with growth or movement.

Think of yourself as a cell.

As you have skin, cells too have outer, protective boundaries; cells are separate and distinct from their environments. Single-celled organisms like those the monastics grow have a cell wall surrounding a cell membrane. Other cells, like ours, have only cell membranes, but the same basic principle holds: this separation allows the maintenance of things inside and the control of what goes in and out.

Human cells also have membrane-bound compartments *within* themselves. These are analogous to organs within a body, and so are called organelles. Each organelle has its own functions, made easier to regulate because of its encompassing membrane (figure 1.3).

How might information pass in and out of these boundary membranes, and what might that information look like? What are the "eyes and ears" of cells?

Cells, the monastics decide, must have "sensors" in their walls or membranes that can interact with the environment. As the monks and nuns develop such concepts, we teachers fill in the details. Cells do indeed have such sensors; they are usually proteins that specifically recognize and respond to the environmental change they sense.

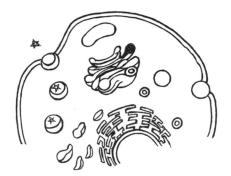

FIGURE 1.3 Some of the membrane-bound organelles within human cells.

How do we get information from inside to outside, *translate* information from one environment to the other?

The key here leads us to another central theme of biology: structure relates directly to function. Change structure, change function. This is true for molecules, cells, organs, and organisms. Ears are shaped to sense and focus sound, fingers to bend and grasp, eyes to capture and interpret light, and cell receptors to sense signals.

On cells, sensors are proteins that change shape in response to signals from the environment. Protein sensors evolved to interact with specific signals. The signal might be an energy molecule such as a sugar that binds to the sensor protein, changing its shape in such a way that it opens like a channel in the wall to let the sugar into the cell. Or the signal might be a temperature or pressure change, either of which can also change the shape and therefore the function of protein receptors.

Signals might also be hormones, short proteins, or just about any other environmental molecules or changes you can imagine. Cells use these signals to translate into responses what is going on around them, to build a kind of awareness.

IN THE MONKS and nuns' bacteria projected on the wall, the signaling molecule, sugar, goes all the way into the cell (figure 1.4). In other cases, signals don't enter the cell, but their binding to receptors and the resulting receptor shape change pass the information that the "signal is present" into the cell. The receptor spans the membrane from outside to inside. Thus, the binding on the outside of the cell changes the shape and function on the inside of the cell.

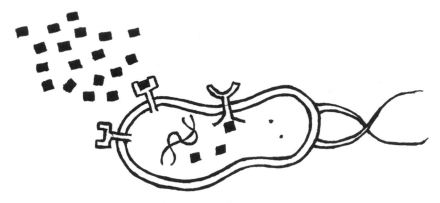

FIGURE 1.4 Sugar (black squares) binding a bacterial receptor
and changing the receptor shape to allow sugar to enter the cells
and cause changes in the behavior of the bacteria.

Once the monks and nuns' sugar gets into the bacteria, it binds to
other proteins, so the message can be passed on to other parts of the
cell. Some of these proteins break down the sugar to use for energy.
Others signal the proteins of the flagella—the hairlike projections the
cells use for movement—so that the bacteria cells move toward the
sugar source. Still other proteins bind sugar and then, as a result, turn
on or off genes on the bacterial chromosome; these genes encode proteins that help respond to the presence of sugar.

Such responder proteins might include more cell surface receptors
that move to the cell surface and bind more sugar, or more proteins
(called enzymes) to digest the sugar, or yet other proteins that use the
resulting energy.

As the monks and nuns learn, proteins are encoded in genes, and
the genes are made of DNA. Upon signaling, genes are turned on or off
when proteins bind to their regulatory regions.

An example of this is shown in figure 1.5 from a mammalian cell in
which the Y-shaped molecules binding the receptor on the outside of
the cell alter the receptor on the inside of the cell so that the dumb-
bell-shaped molecules are released to bind DNA and thus activate genes
to make proteins (the nucleus and its membrane that surround the
DNA are omitted for simplicity). The result of this "gene activation" is
that a series of special proteins read the DNA code of the gene and use
that code to produce a new molecule called RNA (in human cells like

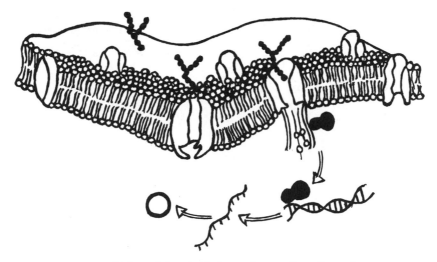

FIGURE 1.5 Y-shaped signals bind receptors on the cell membrane, changing the shape of the receptor and releasing a signal (solid black) that enters the nucleus (not shown) to turn on a specific gene.

the one shown, the DNA is actually in a separate compartment, not shown here, called the nucleus; bacterial cell DNA is not enclosed in a nucleus). This RNA code is then itself read in order to make a protein through a process called, appropriately for our discussion with the monastics, *translation*.

So the translation of information is at the heart of all levels of awareness—from the molecular to whole organisms.

The translation of the fact that the bacteria have sensed sugar happens because proteins directly or indirectly interact with the sugar and change their shape to take on different functions. When enzymes in the cells bind sugar, they change their shape so that they break down the sugar into smaller pieces to make energy. The proteins that turn on genes in response to binding sugar also eventually facilitate the production of more proteins. And the proteins that turn the flagella do so as a result of binding sugar (or another protein that has bound the sugar).

WHEN THE MONKS and nuns add sugar to the bacteria on one edge of their microscope slide, a dramatic change occurs in the bacterial behavior they can see projected on the wall. Bacteria become frenetic, tumbling this way and that. They are sensing the gradient of sugar

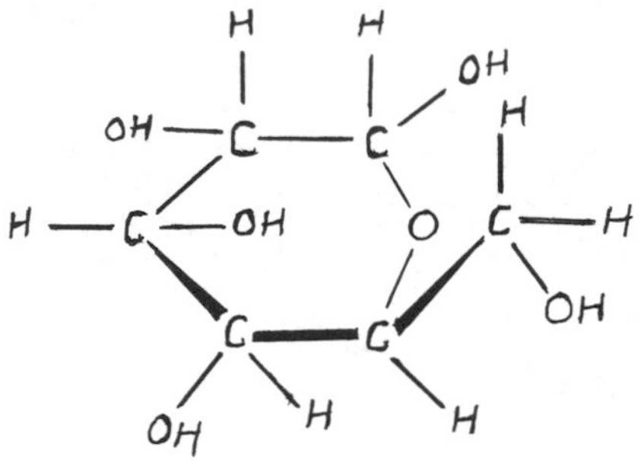

FIGURE 1.6 Molecular structure of glucose: C is carbon; O, oxygen; and H, hydrogen.

and moving toward the source of the sugar, the location of its highest concentration.

Sugars, such as glucose and sucrose, are carbohydrates. This means they contain many bonds between carbons (C) and hydrogens (H), as shown in figure 1.6. Each of these bonds stores energy (think of the bonds as wound-up rubber bands) that enzymes can translate or convert into energy for life in the bacteria. Similar enzymes carry out such energy-producing processes in virtually all cells in all organisms. After all, obtaining and manipulating energy is, at the most basic level, what life is about. The flow of energy through cells and living systems is, then, another of the central themes of biology.

WE WRITE TEXTBOOKS for the monks.

To make our project sustainable, to make it part of the monastic curriculum forever, we need to reach a point where the monks and nuns are teaching themselves, when Konchok moves, as he now has, from being a student to being a translator and a teacher at the same time.

The textbooks are one step in this direction. Geshe Dadul (*geshe* is the title for those monks who have completed the highest degree in monastic studies) was translating our textbook on cells.

Dadul is a diminutive polyglot who can just as easily keep spellbound a class of twenty undergraduates discussing the scientific evidence for reincarnation—people who "die incompletely" (say at a young age, in a tragic and quick accident), recalling details of that past incomplete death and the life leading up to it when they are young children in their following lives, and moles and scars that appear to be passed on from a previous life to these children—as he can outdebate a long-

THE ENLIGHTENED GENE

time intensive-care physician in less than a minute, leaving the latter speechless.

Geshe Dadul and I had finished reviewing most of his questions about the translation, but he had one last one: "In this book, you write about cells doing this and cells doing that; are you talking about cell *bodies* or cell *minds?*"

This is the kind of out-of-this-culture question not unusual in our project, the kind I don't even know where to begin answering. We go back and forth, and I realize Geshe Dadul's question is actually about cell sentience.

In the West, the question of whether a cell is just a collection of parts, merely "a body," or if it has crossed a threshold into being "a mind" or being sentient, rarely if ever occurs; however, in Tibetan Buddhism the question is an important one—so important that it's *built into the language.*

When discussing any sentient being in Tibetan, you add either a body or mind suffix, depending on your meaning. For example, when we talk about studying traits of an organism—a fly or a mouse—Dadul would add the body suffix to the name of that organism, whereas in discussing the intentions or goals of an organism, he would add the mind suffix.

"We scientists don't usually think of cells as having minds," I respond, a bit unsure of my footing here. "Perhaps many cells together could make a mind, but not one alone."

Then Geshe Dadul asks: "But what about *single-celled* organisms like bacteria? Are they cell minds? Is a cell alone sentient, or does it have to be in a multicelled context?"

Should, then, our translated text always refer to cells as bodies, or sometimes as minds, depending on the context? Should the cultural differences that led to the questions be addressed in the text? Should we try to translate the cultural differences themselves?

In the end, we decide that because the texts are telling the Western science story from the Western science perspective—although with examples from Buddhism and monastic life woven in—we should leave the cells without minds, the way Western scientists tend to think of them.

But what will the monks and nuns say to this after analyzing their bacteria-sensing experiments?

At first I thought cells were just rudimentary pieces, like bricks; now I realize each cell is a universe.

With more time, knowledge, and context, Konchok and the other monastics begin to think that a cell is a universe unto itself—if not sentient, then certainly more than just a collection of parts.

An emergent property is when things combined equal more than the sum of their parts. The brain is just a bunch of cells, but when put together in just the right way, a mind emerges. We will interrogate this concept of emergent properties—its power and limitations, its interaction with Buddhism and science—throughout the book.

The separate, individual members of the orchestra sound good, but nothing like the music that emerges from a full orchestra belting out Beethoven's Fifth. Hydrogen and oxygen have distinct properties on their own, but neither has anything at all like the properties of the water, H_2O, they combine to form.

Under everyday conditions, hydrogen is a colorless gas that is not very reactive and exists as H_2, two atoms of hydrogen bound together. Similarly, under normal conditions, oxygen is a colorless gas that exists as O_2. Unlike hydrogen, oxygen is very reactive. When these two elements combine to form water, a unique substance with unique emergent properties is formed.

The characteristics of water allow life as we know it to exist. When we search for life on other planets, the first thing we check for is a sign of water. Most of a cell is water, most of our bodies is water. Water can exist easily as solid, liquid, or gas. Unlike many other substances, the solid form of water, ice, does not sink when placed in its liquid form. If it did, the earth's bodies of water would have frozen solid during past ice ages, and life would not have had any of the liquid water needed to exist and evolve.

Water has unique chemical bonds that allow many other chemicals to easily dissolve in it, chemicals necessary for life. Water absorbs huge amounts of heat, so the oceans and lakes help regulate earth's temperature. The ability of water molecules to stick together allows water to move up through stems to nourish plants, many of which nourish humans and other organisms.

Similarly, the different parts of a cell can function independently; but together they become the stunningly complex fundamental unit of all life on earth, a functioning unit that can spin and dance, sense and move toward sugar added to a microscope slide, or send signals among themselves in a brain so you can read this book.

Within cells there is a very systematic world—lots of things are going on . . . signaling, making energy. It's really sort of strange. When you have many of these cells, they come together to form tissues and organs. So we see emergent properties.

In Tibetan thinking, blood, flesh, and water come together to make body, mind, and emotions. It seems as if both Buddhism and science have similar ideas: that every bit of matter starts from simple and develops into complex. Things that arise are interdependent and have emergent properties. For example, the human physical body is comprised of many organs and parts that alone are not a body but together are.

Buddhists believe that our physical body is formed from a combination of an egg and a sperm. Egg and sperm themselves are made of five elements. At the subtle level, the five elements, called earth, water, fire, wind, and space, are the basic building blocks of our bodies. The combination of these five elements form flesh, bone, and blood. The combinations of flesh, bone, and blood form organs, then organisms, then populations.

Biology is the study of the interactions of molecules—DNA, genes, proteins, enzymes, and their functions—that develop into all the different tissues in living organisms. In biology, "emergent property" means that things are formed from a combination of multiple causes and conditions, and gradually the things become more and more complex.

We are talking about what I call the Living Staircase. It's a conceptual framework that a colleague and I developed for biological knowledge that I describe to the monks and nuns and to my students back in the United States.[1]

The Living Staircase refers to the different levels of life that build upon each other, moving from atoms to molecules, to cells, to tissues, to organs, to organ systems, to organisms, to populations, communities, and ecosystems. Another way to think of an emergent property, as Konchok points out, is one that emerges from a combination of characteristics, functions, or constituents at one level of the staircase and often leads to the next level.

Hydrogen and oxygen *atoms* combine to form water *molecules* with characteristics much different from either atom alone. The organic *molecules*—proteins, fats, sugars, RNA, and DNA—combine to form a *cell*, a universe profoundly more complex and with more possibilities than any of the molecules on their own.

Liver cells combine to form an organ with new characteristics greater than the sum of those of the cells alone. As will become evident, all the central biological themes—structure and function; evolution, conservation and diversity; communication and information flow; homeostasis, cycling, and regulation; energy flow; interdependence; and division of labor—operate on each level of the staircase.

TAKING A WALK on the staircase, so to speak, helps the monks and nuns see the broader context.

In their bacteria-sensing experiment and discussions of cell biology, they learn that the single cells they isolate and grow interact with and respond to their environment. All organisms send and receive signals. Communication is a universal theme of the Living Staircase.

Take, for example, the communication occurring in our classroom.

Each translator is different and affects what and how the monks and nuns hear and learn the ideas and information being taught. Tsondue is among the best. He knows the most languages and the most subtleties, turns of phrase, and metaphors. Tsondue is Tibetan, studied to be a monk in India, and then came to the United States and earned an undergraduate degree in physics. He moves words and ideas back and forth among languages and cultures with enviable ease.

This translation, teaching, and learning—processes requiring high-level sentience—are communication at the *organism* level of the staircase. Down a few steps, the act of teaching involves a collaborative communication among several *organ systems*: the nervous, muscular, skeletal, and respiratory systems.

Each of these systems is composed of a group of specific organs. The vocal folds of the teacher generate sound by opening and closing over a stream of air coming from the lungs and moving through the larynx. This process is moderated by the nervous system, specifically the brain thinking words and signaling them to the vagus nerve. And while the teacher breathes in and out, the muscles and cartilage of the larynx open and close the vocal folds to help produce spoken words.

At the same time, Tsondue hears my words. Hearing is a neurological process that involves much inter- and intracellular communication. In this case, my larynx produces sound. As the monks and nuns learn in physics, sound is a vibrating wave with mechanical energy. The sound waves enter Tsondue's ears, and in the inner ear the waves push at small hair cells (the *cellular* level of the staircase). The pushing on hair cells forces open protein channels (now at the *molecular* level of the staircase) in the cells' membranes, channels similar to those on the cell surfaces of the bacteria the monks and nuns are studying.

Hair-cell channels allow the movement of small charged particles (ions, now at the *atomic* level of the staircase) across the cell membrane, in the process converting the mechanical energy of sound into the electrical energy of a nerve signal. The nerve signals communicate with the auditory nerves of the ears and then with the brain. In Tsondue's brain, complex neural communication among groups of neurons allows translation of what he heard me say. Then Tsondue talks via the same processes I did, but in Tibetan instead of English.

And so, for teaching to work—in order for this *interorganism* communication to occur, communication must occur all the way down the staircase within organ systems, organs, tissues, cells, and molecules.

At first the Living Staircase analogy might sound hierarchical, linear, and unidirectional, but at any one level, evolving emergent properties that affect and shape the steps *above* it also potentially affect and change the steps *below* it, which then will circle back and affect itself. Similarly, each step, rather than being independent from the others, *encompasses* and *includes* those below it.

NOW A twist.

We usually think of bacteria as solo cells doing their own thing (often, we imagine, trying to infect us).

What if bacteria could act like multicelled organisms and blur the lines between the cell and tissue steps of the staircase? What if they could cooperate with other cells of their type to share the news, say, that there is sugar on the microscope slide? What if they could share resources with each other, cooperate? Would this have the monks and nuns leaning toward calling bacteria sentient?

And more: What if bacteria could cooperate with *other* species' cells? With human cells? Imagine if we humans, apparently the ultimate sentient beings (at least in our Western minds), *require* bacteria for our *own* sentience?

Bacteria actually have a rich social life—both with cells of their own species and with cells of other species. Signaling among and between cells and organisms, even so-called single-celled organisms, is vital for life. A central theme of biology is communication; virtually all life is communal.

Microbial life can exist in structured communities called biofilms. To establish and maintain these many-celled structures, bacteria use quorum sensing—cell-to-cell communication among their own species. This uses the same basic kind of signal response as the bacterial response to sugar, except in this case the bacteria release the signals themselves (as opposed to having the signal added from an external source like that provided by the monks and nuns).

More striking, many bacteria use quorum sensing when food supplies are low to gather themselves into complex multicelled structures that allow them to save their next generation of cells in protective spores until conditions improve. All of the other cells in these structures sacrifice themselves for the good of the spore cells—bacterial altruism.

While the actual physical DNA of each cell is not passed on, the genes of the bacteria that sacrifice themselves are passed on through their close relatives. Within their biofilms some bacteria act like electrical wires, transferring energy in the form of electrons among themselves, so all cells in the biofilm can benefit from the electron energy to which only some of them have direct access. This phenomenon appears to occur even between different species of bacteria.[2]

These communal activities represent evolutionary steps from single-celled to multicellular organisms. For example, in spore-carrying structures, some cells that at one time had the same potential as the others change their personalities and commit for the rest of their lives to new, distinct personalities. This is exactly what multicellular organisms are —collections of cells with the same potential that commit to being particular types of cells with very different personalities doing very different jobs for the good of the whole organism.

Bacterial cells are us; bacteria live in and with us and affect human behaviors, help us digest food, and influence our neural and immune-system development. In turn, humans provide for the bacteria a safe place to live and reproduce.

Humans and many microorganisms (fungi and viruses, in addition to bacteria) have evolved not only to tolerate each other, but also to *share* information and *cooperate* with each other — if not a sentient relationship, then certainly a very Buddhist-sounding kind of relationship. The sum of these "good" microbes that partner with human cells is known as the microbiome. We have ten times more bacterial cells living in us than we do human cells, and these bacteria have a hundred times more total genetic information than our cells do.

Pathogenic bacteria, viruses, and other microorganisms also drive evolution. If humans had not evolved and do not continue to evolve to fight them, we would be dead or sick at best. The problem is that scientists discovered the existence of the "bad," pathogenic bacteria first, a couple hundred years before discovering the "good" bacteria of human microbiomes.

And it was shown, in systematic and scientific fashion, that these microorganisms were killing people en masse. So the second half of the twentieth century was in large part devoted, in medical terms, to *killing* bacteria. Unfortunately, and unappreciated at the time, those beneficial bugs of human microbiomes were also simultaneously being killed. This lesson, of using innovative technologies to solve sociological and ecological problems while ignoring or being unaware of potential side effects, is something science has experienced repeatedly.

The twenty-first century has seen the microbiome become a star. Some scientists have gotten a bit breathless about the role, effects, and importance of the microbiome, but it's clear that a new vista has opened on our understanding of how our bodies work. Just as a new technology — the microscope — first brought microbes into view, very refined DNA sequencing technologies have allowed the uncovering of the new world of the microbiome. If the transformation in knowledge and thinking brought about by the latter has not been quite as dramatic as it was with the former, it has sure been close.

Is it a stretch to say our microbiome affects, or even is integral to, our own awareness or sentience? Our gut microbiome definitely talks to our brains. Bacteria that inhabit mammalian guts can produce signaling molecules chemically identical to the ones mammals produce. One of the major molecules made by mammals that inhibits signaling in the human central nervous system (the physiological core of our sentience) is also made by one of our gut bacteria. This molecule decreases pain and inflammation and so might be important in treating certain human diseases. Other symbiotic bacteria produce and release acetylcholine, a signaling molecule of the human nervous system involved in cognition and attention.[3]

Because human and bacterial cells share the same basic cellular parts and strategies, the two types of cells are able to communicate with and help each other. The microbiome is part of a vital three-way conversation with the immune system and the nervous system. In experimental mammals, the amount and types of species of bacteria in our microbiomes alter with stress (say, lack of sleep or viral infection), and stress can be relieved by artificially readjusting those microbiomes back to prestress states.[4] Good bacteria appear to affect levels of bad bacteria, thus contributing to rates of infection. Artificial killing of neurons in mice alters their microbiomes. Microbiomes affect the development of immune systems, and immune-system molecules are *required* for learning.[5]

In chapter 7, "Meditation and the 'New' Diseases," we'll unpack more of the biology of this conversation. For now, the relevant point is that the original question with which we challenged the monks—are bacteria sentient?—becomes more complicated. Maybe bacteria *are* cell minds. Maybe they're *our* minds. Perhaps if we'd thought of them as sentient in the first place, the nature of our science and its results would have been better. It definitely would have been different.

ONCE CELLS START collaborating, their capacity for awareness increases greatly, and surely this provides a great survival advantage. As with two humans, once two cells work together, they can share responsibilities. Such division of labor occurs *within* each cell—for example, information is carried and passed on by chromosomes, and membranes protect and separate the internal organelles. But division of labor becomes particularly advantageous at the next levels of biological organization, when jobs are shared *among* cells.

Symbiotic relationships—such as the microbiome-human one just discussed—provide some of the most astonishing examples of division of labor. More and more such cases of symbioses between species millions of years apart in evolution are being discovered and probably are the rule, rather than the exception. Aphids, tiny insects that eat plants, cannot live without a particular bacterial species that inhabits their guts, and *that* species cannot live without aphids.[6]

One type of squid relies on a particular species of bacteria to illuminate its body at night while it hunts in the ocean. These glowing bacteria alter the moon shadow of the squid on the ocean floor to fool its predators; if squid embryos do not acquire the bacteria, they die, because the bacteria, in addition to preventing predation of the squid, also signal the squid's developmental maturation. No bacteria, no adult squid.[7]

And mitochondria, the organelles that produce energy in most all plant and animal cells, are thought to have once been separate bacterial cells that entered into a mutually beneficial relationship with "host" cells and then stayed for good.[8]

COMPASSION, ALTRUISM, and empathy are at the very core of Buddhism. By definition, these attributes require social interaction among humans and other sentient beings. Such positive social behaviors—whether among single cells or humans—increase survival, decrease suffering, and often increase quality of life. If you are lonely, this is reflected in your mental and physical health. Lonely people tend to take on more bad habits, suffer more commonly from diabetes or mental illness, release higher levels of stress hormones, and thus have shorter life spans than socially engaged people. As we will dive into in more detail, then, natural selection favors the evolution of social behaviors —in *all* organisms. The more social, the more sentient (and vice versa).

OUR PROJECT involves a different kind of symbiosis that I do not see at first.

It's late afternoon, and I'm strolling in the foothills with Dhondup, one of the monks who is already fluent in English. The air is electric;

we've just been up at the tree line, and as I learn later, some of the monks hadn't been in the high-altitude mountain air of their youth since they first escaped from Tibet and traveled to the monasteries in the lowlands of south India years before. They're giddy.

Dhondup is a big basketball fan and a devoted monk. I ask him why he is doing this, why, instead of focusing on his monastic studies, is he here participating in our somewhat wild, pie-in-the-sky experiment, learning modern science far from his home monastery?

First, he says, because the Dalai Lama thinks it's a good idea. Good answer. And then:

"I study modern science to understand my Buddhism better."

This pulls me up short. I literally stop walking. This is the symbiosis I hadn't seen. The insight itself sounds so un-Western to me. Our project isn't about converting people—either Westerners or monks—but rather about enhancing and deepening the knowledge our students *and we* already have.

I return to Dhondup's foothills insight often. Wherever I am teaching—whether in India or the United States—this is the great challenge of teaching, one I hadn't fully appreciated.

WE HAVE DISCUSSED "*why* complexity." Why sentience? The other part of the question is "*how*."

Consider the bacterial cells in the monks and nuns' sugar experiment or cells in the human brain. Human cells (as well as bacterial cells within any one given species) have virtually the same genetic information and potential capacity, but use that capacity differently *depending on their location in time and space.*

What a cell is at any given moment is a snapshot of the ongoing conversation between its genes and the environment—awareness at its most fundamental level.

In biological terms, think of time and space as dimensions of the environment in which cells exist and of which they are aware. Biologists consider distinct timeframes: circadian time—spanning about twenty-four hours; developmental time—spanning hours to months, depending on how long it takes the organism to reach biological maturity; and evolutionary time—spanning generations.

Using the monks-as-cells analogy again, within the circadian time frame, monks respond to sunlight and have sleep and wake cycles. Within the developmental time frame, they developed from the single cell of a fertilized egg and continue to age and change. And considered from an evolutionary time frame, they received their genetic inheritance from both parents, who received it from their parents, and so on back in time.

Similarly, the bacteria studied by the monks and nuns probably behave differently—meaning, they respond differently to the environment by regulating different genes—depending on the *time of day*. This relates to how much light they or their food sources require.

We do not usually think of bacteria across developmental time, but they do change during their lives, like the bacteria mentioned previously that under starvation conditions convert into a multicellular spore structure. Also, bacteria gather and share and change information in other ways: bacterial DNA can mutate like human DNA; bacteria give and accept new genetic material to and from other bacteria; and they actually have immune systems that "remember" viruses that infect them, so that they can better fight that virus the next time it attacks.

Across evolutionary time, because bacteria can mutate and can reproduce a new generation in minutes, they are able to adapt and evolve in days or even hours in response to environmental change. One bacterium, given the right conditions, can multiply into a billion in a few hours. So a bacterium (which has a version of a gene that happens to be altered in such a way that it gives that cell an advantage) can rapidly predominate in a particular environment—the advantage of relative simplicity. Perhaps one reason bacteria have evolved to assist in so many important human biological functions is exactly because they can adapt and change so much more rapidly than our own cells can.

Human cells are intriguingly similar to and different from bacteria in the time dimension. Human cells' response to circadian rhythms, the daily rhythm of light and dark, can be dramatic. Similarly, as we explore in the next chapter, the same set of genes can respond in diverse ways to signals from the environment across developmental time, resulting in the development of strikingly different cells.

Beginning with the division of the fertilized egg, each individual cell is instantly in a unique environment. Each has more or fewer neighbors of one kind of cell or the other and is closer or farther away from

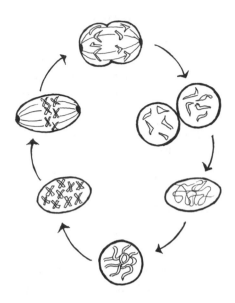

FIGURE 1.7 The cell cycle, with chromosomes indicated.

particular environmental signals. Based on these different environments during development, each cell receives and responds to different signals, differentiating itself into one part of the division of labor story, eventually resulting in a complex multicellular human. In the end, the nerve cells of the brain look and act very different from muscle cells, and this is reflected in and a result of how many of which proteins and other cellular constituents they contain and when.

Cells also operate within a fourth time frame: their own individual growth and division cycles (figure 1.7 shows the division, or mitosis, of a typical cell, with chromosomes highlighted). Cells that have access to an energy source, such as the monks and nuns' bacteria on their homemade food plates, are often busy growing and then dividing to make two new cells just like them. In multicellular organisms, some cells, like many brain cells, get to a certain point in life and stop dividing; others kill or sacrifice themselves; still others—such as the stem cells to be explored later—divide to make two different kinds of cells from one.

The capacity of a cell to divide into two clones of itself is challenging enough for this discussion; a newly made, exact copy of genetic material is required, as is two cells' worth of all the other parts of a cell. While a bacterium's job of cloning its one small chromosome of genetic material is relatively easy, imagine producing an exact copy of all 3.1 billion nucleotides of the human chromosomes every time the cell divides. And some human cells—such as skin or hair cells—divide constantly. Where a cell is in its cycle of growth and division is important to consider.

Many senior monks think if we bring science ideas into the monastery, young monks might change their minds and neglect Buddhism.

For me personally, when I started learning science before, and now looking back, it's not just learning science; it also helps me reflect on my own philosophy.

Before I started learning science, I learned about reincarnation, creation; I read texts. Now I look at these things more deeply. In science you can do experiments to check; there is evidence. But in oral philosophy there is not this kind of evidence; some other monks won't like this.

In my mind, I am always asking: What will be in the next life? How did universe formation happen? But these are mysteries.

Life is complicated. We generalize about what "scientists" or what "Buddhists" do or believe, but of course, it's never that easy. Just as many different ideas and opinions exist among scientists as exist among Buddhists.

Many different types of Buddhism exist; there are even numerous types of Tibetan Buddhism. Although our science-teaching project was the Dalai Lama's idea, and he promoted it, acceptance of the project among monks and nuns has taken some time and still, years since it began, varies from monastery to monastery *within even the Dalai Lama's own sect of Tibetan Buddhism.* Some of this resistance is expected: a fear or distrust of the new, a concern that monasteries might lose even more monks than they already do to the secular life. Another piece of this resistance is that Tibetans associate science with the scient*ism* promoted by communism and China—the nation oppressing Tibetan culture.

During a lunchtime break in Dharamsala, as I attempt to return the impossibly spinning ping-pong shots of a monk visiting the college for the weekend (apparently, during the long monsoon seasons, some monks have enough downtime to learn to make ping-pong balls do things I'd never before witnessed), I realize this: it is the very differences and tensions *within* Western science and *within* Tibetan Buddhism that, together, may allow for the most striking new ideas to emerge from our project. Identify the tensions and work with them. I lose 21–2 only because ping-pong monk tries very hard and humbly to allow me to hit the ball once or twice.

ARE THE CELLS dancing on the laboratory wall and moving toward sugar *aware?*

We ask the monks and nuns to debate the question. After days of exploring its biology, the monastics attempt to settle the sentience question using their age-old approach to dissecting complex issues.

Tibetan monastic debates are not at all like American debates. Clearly, monastic institutions long ago came to appreciate the importance of involving in learning as much of the body and as many of its senses and cells as possible. Maroon-clad men yell fiercely, grab and rip at each other's clothes, team up and whirl like dervishes, slap their own hands together horizontally with a pop for exclamation. Sweat pours off their bodies as the monks make philosophical points, bellowing at the judges, scrapping for an advantage. As soon as you signal that the debate is over, as if turning a switch, the monks immediately return to their quiet, relaxed selves.

This seemingly unmonklike display, this debate, is a common and powerful way for Tibetan Buddhist monks to communicate, challenge each other's knowledge, and even take their exams. It is physical and mental exercise all at once; it is strategizing and teamwork, a friendly but aggressive way to work through conflict and tension, mixing sport and intellect.

We divide the class into two teams; they prepare their arguments and then debate, literally wrestling with the sentience question we have been exploring for days. The already hot classroom heats up even more. There is an occasional pause so that the rapid, emotive Tibetan can be translated for us teachers.

Following the debate, as we gather for the last cell biology teaching session, we ask the monks and nuns to raise their hands.

How many think bacteria are sentient beings?

Half vote yes, half no.[9]

CHAPTER TWO

Life, Death, and Sacrifice

I stare at slime molds—and they're beautiful.
JOHN BONNER

*A tribe including many members who, from possessing in a
high degree the spirit of patriotism, fidelity, obedience, courage,
and sympathy, were always ready to aid one another, and to
sacrifice themselves for the common good, would be victorious
over most other tribes; and this would be natural selection.*
CHARLES DARWIN, *THE DESCENT OF MAN*

eaching Tibetan Buddhist monks fundamentally changed the way I think about science, about how easily we can be fooled that "the science we know" is right; how deeply our science is affected by our view of life, our societal and personal philosophy; and how this profoundly affects the questions science asks, who asks them, and thus what science is done by whom and to what ends. This became dramatically evident when teaching Konchok and his friends the most basic processes of biology—development from sperm and egg to adult. Buddhism changed *everything*.

In this chapter, we show how. We revisit the processes of development through our experiences teaching them to the monks and nuns, but dig into the crevices of the field and uncover knowledge that is there but is hidden or deemphasized when development is taught and understood in the West, knowledge that became evident and uncoverable only as a philosophic whole when teaching and learning from the monastics. When I learned or taught biology previously, what we focus on here was glossed over or entirely omitted, *because it does not fit the classic Western philosophic-scientific narrative.*

From the moment we begin to live, we begin to die. Death is just as exquisitely regulated as life, and the former is absolutely required for the latter. And yet in the West, our focus is on life and living and avoiding death—both in the real world and in the classrooms and laboratories. What if death is considered, not as an end unto itself to be avoided, but rather, in the way that Buddhists think, as a sacrifice or as part of a cycle of new life to come?

Even before the egg that became you met the sperm that would fertilize it, cells sacrificed themselves left and right toward this end. Men produce sperm prodigiously, as many as one thousand per second. Programmed cell death—apoptosis—occurs all along the pathway that produces sperm, way back to when your father was a fetus inside *his* mother. At that time a wave of cell death occurred in order to reach just the right ratio of developing sperm to cells that would eventually nurture those sperm. At this stage and throughout life, regulated cell death also removes faulty sperm, especially sperm with damaged DNA. As many as three-quarters of developing sperm kill themselves before maturing. Such death vastly increases the chances that high-integrity life will occur in the next generation.

Another big wave of sperm death occurs after sex, in the female reproductive tract. Tens of millions of sperm enter, but only one fertilizes the egg. Why so many at the start? Maybe a community effort is necessary; so many sperm may be present to ensure that at least one makes it to the egg. One theory is that inside the female, on their way to the egg, different populations of sperm are "activated" at different times in her reproductive tract to help navigate the challenges therein—thick mucus and the female's immune system, which attacks and kills sperm.

Unlike fathers and their sperm, mothers are born with virtually all the eggs they will ever have. The egg that became you was one of over twenty thousand pre-egg cells in your mother when she was developing inside your *grand*mother. *Greater than 90 percent* of those pre-egg cells died somewhere along the line from the time they were pre-eggs inside your mother when she was an embryo to the time soon after her birth, and this death probably occurred in a carefully regulated fashion to allow for only a small percentage of those eggs, and only healthy ones, to survive.

Think of every cell that lives and then dies as sacrificing itself for the cause. Of course, all our cells die in the end, but billions of others give themselves up along the way to allow us life.

Traditionally in the West, we focus nearly exclusively on the one sperm that makes it and the one egg it fertilizes.

WE RIG UP a projector in the Dharamsala classroom, and the monks and nuns settle onto their cushions; it's still early, the inevitable heat yet to press into the room. The video we'll show is about slime mold.

I look out at the class. They are much like any group of students settling in, some laughing and (I imagine) chatting about the night before, others reading their notes or wondering why they're there at all.

I was up with the sun, as it spread early morning light over the Himalayas and haloed the monks pacing the roofs of the dormitories across the way, praying their way back into another day. After meditation and breakfast, I crossed the courtyard, skirted the basketball court (where young Tibetans play their own unique style of very serious hoops late into the night), and headed up to our classroom.

Sara College is a quiet retreat, a quarter way up the ridiculously narrow and twisting "road" between the chaotic bustle of Indian villages below and the overwhelming yet peaceful crush of the small-village-turned-international-spiritual-center, Dharamsala, above. Sara is one of the colleges the Dalai Lama opened in the diaspora to help Tibetans maintain their language and culture. Tibetans come here from India and other parts of the world for a regular college course of study or return during summers to better understand and maintain their identity.

Dharamsala perches above the college like a metaphor for Tibet and its people—precariously balanced on a mountainside, an earthquake away from obliteration—yet deeply rooted in place by the Dalai Lama's residence and surrounding temple, visitors from every corner of the earth (seeking or merely curious) spinning the prayer wheels, their hundreds of pairs of sojourning shoes lined up outside awaiting the return of their owners.

The Tibetan government in exile is also here in Dharamsala, crammed creatively into one easy-to-miss compound. And one can occasionally view through the haze the stunning snow-covered peaks that once en-

tirely isolated Tibet—on the opposite side—from the rest of the world. Peaks that have since claimed many lives of those seeking escape.

Sara is an odd place to travel to from halfway around the world to teach development, genes, cells, neuroplasticity, and how to "think like a scientist." Our journey, we think at first, is different from the other seekers.

IN THE VIDEO we show the monks and nuns, we see what appear to be single cells minding their own business, milling about like people on a winding village road, each doing their own thing. But suddenly, as if on cue, the individual cells gather together and become one many-celled creature (figure 2.1). Then, even more striking, this creature radically changes its shape and transforms into a stalk with spores on top. What happened?

The answer to that question encapsulates much of developmental biology—the biology of how new organisms are created and become —and we return to this visually stunning performance over and over again throughout the week.

Slime mold are probably not the first organisms that come to mind when you consider beauty or ponder the great questions of life and biology, such as: how is it that I was once one cell and now I am billions of them, living and dying, and working together to read and understand this sentence? But it turns out they are a superb model system for exploring such questions.

Slime mold live on the forest floor as single-celled organisms, happily eating bacteria. When they run out of food or are otherwise stressed, these single cells find each other and come together in clumps to transform into multicellular organisms (similar to those bacteria in the last chapter)—the sluglike creatures the monks and nuns see on the time-lapse video.

The slime-mold life cycle is shown in figure 2.1. Once essentially identical single cells now adopt dramatically different roles; some become stalk cells and others spore cells. The stalk cells raise the spores up off the ground, where they store the next generation until conditions improve. Then spores are released and spread to start the cycle over as single cells—the stalk cells having sacrificed themselves so that their genes can be passed on, via the spores.

From fertilization to birth and from birth to death, an individual lives and dies every moment, leading to life after life.

We Buddhists sort all knowable phenomena into two categories: permanent and impermanent. Impermanent phenomena, such as material objects, are always changing; nothing remains the same. Birth, growth, decay, and death are inevitable for all material objects, people, societies, and states of mind. It is easy to see the constant change that takes place in all things—the birth, the aging, the dying, and again the rebirth.

Right after birth, all of us are gradually growing old, and our entire lives we undergo constant internal and external changes; some of these changes involve decay and change into different forms, and some cells and organisms die as a sacrifice for the benefit of others.

Buddhists say dying is natural and is mainly caused by aging and conditions such as disease, outside violence, trauma, or suicide.

All beings suffer in Samsara and, therefore, people practice meditation for the purpose of ending the cycle of rebirth and being released from suffering. The idea is that birth leads to aging, aging leads to sickness, and sickness leads to death. From the biological perspective, it is true that most developmental processes involve as much death as they do life.

Many organisms such as humans and animals sacrifice their lives to help others and allow others to stay alive and ensure the successful continuation of future generations. Self-sacrifice or altruism is a core virtue in Buddhism, and it is mainly based on ethics and morals.

Most Americans think science is right; it's so deep in our culture as to be an unconscious part of who we are. Even hard-liners, who consider evolution anathema and don't trust the climatologists' claims of doom, in practice "think like a scientist." Life is linear—A leads to B leads to C, logically and usually in a repeatable fashion. To a large extent we trust our doctors, who we believe use evidence-based diagnostics to prescribe the drugs we take without much question. We assume those drugs will make us better because they have been "scientifically tested." Early in our project teaching monks and nuns and bringing what I was learning from it to Emory, I began teaching a course to undergraduates, "Science and the Nature of Evidence." In the course, we explore the question of why we believe what we do and the role science plays in those beliefs. Virtually every student enters the course believing (they discover after discussion, reading, and self-reflection) that science is right. When they begin to dissect this assumption and see its implications, they get nervous and worried, then excited, and then the real work begins.

The Dalai Lama expressed to us from the beginning of our project that the uncovering of assumptions was a goal of his for both the Tibetan monastics *and* us scientists. He imagined not only that the monastics would learn science and then return to India (and maybe even Tibet) and teach it to other monastics, but also that the monastics would teach and share their ideas with Western scientists and professors—to contribute the Buddhist points of view to the Western community that might help scientists both in their science and in their education. Might slime mold and developmental biology, the stunning dance of life becoming, be a case in point?

Think about the impact of linear thinking on our actions, on how we live, on where we focus our resources and time—especially in science and health. We are born, grow old, and die; we are not reborn. Death is an ending—whether or not you believe in an afterlife. This thinking shapes our lives and our deaths; we must stay young as long as possible, we must avoid death as long as possible, avoid the end. This approach affects even how, and if, we address the most basic questions of science and life.

André Gide said, "Man cannot discover new oceans unless he has the courage to lose sight of the shore."

The main disconnections between Western science and Tibetan monastic traditions are the differences in perceptions, conceptions, and beliefs, and the lack of intercultural awareness.

When not well aware of the essence of another culture, it is often the case that one finds nearly every process problematic. When studying science as a monk, it is much more effective for me to avoid mixing scientific concepts and Buddhist concepts too much, to lose sight of the shore to learn more deeply and more easily unveil personal dilemmas. I do not ignore or lose the Buddhist concepts, but temporarily reserve them.

As His Holiness the Dalai Lama tells us, "Listen to others carefully and then share your notions." So this new approach, this new way of looking, may provide new oceans of wisdom.

Each mitosis, each cell division—the very essence of life, one cell dividing into two others—is, as we learn beginning in grade school, one mother cell dividing into two daughter cells, two new lives. But this most fundamental step of life *is also a death*, a sacrifice: the mother cell sacrifices herself to make two new lives (refer back to figure 1.7). This difference in how the very same phenomenon is viewed matters.

Our Tibetan Buddhist students have evolved a different life philosophy. As we teach them, the monastics learn developmental biology from a different cultural framework, and thus become different kinds of developmental biologists than our students back in America. For Buddhists see life and death as a circle of renewal, each an integral part of the other. Samsara is the continual cycle of birth, sickness, death, and rebirth, a cyclical existence. Except in those exceedingly rare cases of enlightenment, all sentient beings continue in this circle. Integral to Samsara are the core Buddhist virtues of self-sacrifice and altruism.

Demonstrate compassion and sacrifice for others not able to ensure their own survival; be generous to those who have nothing. Give your time, energy, and life for the greater good; sacrifice your life for the community.

The controversial practice of self-immolation is an extreme form of sacrifice arising from the Buddhist culture. Konchok sees self-immolation, in select cases, as a form of sacrifice; for example, when many Tibetans in recent years have self-immolated for humankind and community, in a dramatic call to save and protect their culture, tradition, and ethnicity —a call for freedom.

TO GET A SENSE of what kind of biologists the monks and nuns might become and see how it changes the thoughts and the science of our biological becoming, let's follow the arc of biological development with Western scientific knowledge, but in the context and as part of Samsara, an eternal cycle of life, death, and sacrifice; life is part of death, leading to death and new life.

As our lives and those of other organisms cycle, so do the lives of our cells. Down at the next level of the Living Staircase—that metaphorical construct we described in chapter 1 that moves from atoms and molecules to cells, tissues, organs, organisms, populations, and then ecosystems—biological development is also a balanced cycle of *cells* living, proliferating, and dying. Some cells are irreplaceable, while some live a complete life and then die. Such death, as in the case of the slime-mold-stalk cells, can be thought of as a sacrifice, a kind of altruism. All cells of multicellular organisms are, in a sense, altruistic. They divide labor and work together to keep each other, and their organisms as a whole, alive. (We also saw evidence of altruism in chapter 1, where bacterial cells with access to energy sources shared that energy with other bacteria lacking such access.)

In one of our visits with the Dalai Lama, he spoke on the issue of sacrifice versus selfishness. Is there such a thing as pure altruism without some expected benefit to the self? He responded, "Be wise-selfish, not foolish-selfish." While slime-mold cells clearly do not have the capacity to be selfish in a human sense, perhaps in dying they are being "wise-selfish" because their death allows their genes to continue in the cycle of life.

ONE MINUTE slime-mold cells live as apparently identical, but independent, single-celled creatures crawling about searching for, and then consuming, bacteria for breakfast. How do they become different? Slime mold are easy-to-grow cells that develop and evolve in front of our eyes in real time.[1] This organism routinely (and in a way that can be controlled experimentally) takes the first seemingly magic leap of development—transformation from identical to diverse cell personalities, and from unicellularity to multicellularity.

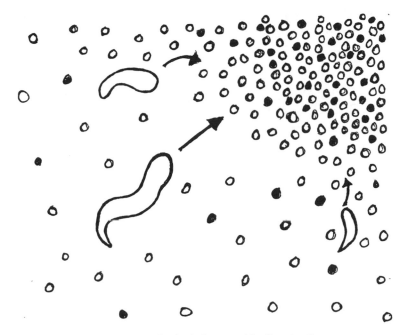

FIGURE 2.2 Individual slime-mold cells orienting
and moving toward an attractant.

When fat and happy and living large, each slime-mold cell makes more copies of itself by dividing into two identical cells. This process of mitosis, as we discussed in the last chapter, is conserved in virtually all cells on earth: the goal is to double all the genetic material in one cell, as well as much of the other vital cell parts, and then divide, so each daughter cell gets a full complement of identical genes. Remember that the mother-daughter cell language is a bit deceiving, as in each cell division, the "mother" cell sacrifices herself and *becomes* her two daughters; the mother no longer exists.

Now another kind of sacrifice occurs. If the slime mold's food runs out—perhaps by dinnertime—these cells undergo a dramatic personality change. The cells begin to live out in *real time* the transformation from single-celled to multicelled. The previously independent and separate single cells send out a signal to surrounding cells: things do not look too good; we need to get together and save ourselves.

Individual slime-mold cells follow this warning signal to its source (in figure 2.2 the source would be in the upper-right corner). The slime-mold cells detect the signal and the direction from which it comes and translate it into movement toward the signal source—like bacteria to sugar. Already, even at this time before the cells have adopted entirely different personalities, discrete differences are taking shape among these communicating cells—differences that set the stage for later changes. One cell is the first to release the warning signal, and others respond to it; some responders are closer to the source than others, so they reach the source sooner. *Differentiation* is happening.

THE SINGLE slime-mold cells were happily grubbing along the forest floor as genetically identical. Initially, they were the same, but then things changed.

Hundreds of monks are together—all more or less the same and doing the same thing: praying in the temple. At the end of the service, one monk at the back door of the temple yells, "I need fifty of you to help me build a shelter behind the temple." The signal is his voice; it creates a gradient of sound. Those nearest to the yelling monk hear him best, respond sooner, and reach him sooner. Those farther away do not hear him as well and respond more slowly. Those at the other side of the temple may not hear the yelling monk at all and therefore do not respond to him. The responding monks orient to the source of the sound and translate the signal into movement down the sound gradient toward him.

The fifty-one monks who leave the temple are now different in important ways—different than they were while they were all in the temple and praying, when they were doing the same thing, dressed in the same way. The yelling monk is the source of the signal and perhaps also of future information. The other monks following him are arranged in space roughly based on where they were originally in the temple. The closer ones reached him sooner in time and thus are now closer to him, those farther now were farther away then. When they reach the site of construction, the monks are assigned tasks in the order they arrive. Now they become even more different, as division of labor takes place. Some work together to carry wood, some to find nails, some to lay a foundation, some to oversee the work, some to bring water. Each monk has the *capacity* to do any of these tasks, but because of where they happened to be in time and space in relation to the initial signal, they now are assigned and take on different roles.

People are not slime molds, and analogies can only go so far. But both illustrate basic principles of development—adhered to by slime molds and all multicellular organisms. Differentiation—what each cell becomes and does—depends on time, space, and signaling; each cell is in a unique environment, so that each cell's genes are affected differently than others, resulting in different functions and behaviors. These functions and behaviors are all in relation to gradients established way back at the initiation of events. Cells coordinate movements and cooperate with each other.

Unlike the slime-mold cells, the monks in our analogy can, after helping to build or at any point they like, change jobs or functions. For the most part, once cells start down the path to differentiation, they complete it and their new "personalities" or functions become determined; that is, they maintain their final state. Many such cells can still divide and grow, but their daughters also maintain the same personality. Stem cells are an important exception to such a life history of cell division, differentiation, and determination, and we will discuss them later.

I took a course on animal behavior at Emory, and that course affected me a great deal. Animal behavior is about species' lives and survival, feeling and emotion, wisdom and intelligence, skills and tools, language and signals, relationships, ways of coping with enemies, cues used during migration, creating houses for living, courtship displays, alarm calls, and feeding and protecting infants from danger.

The impact of this and my other courses is that whenever I see any kind of animal I am thinking about how they should be, their differences with others, their color differences, their appearances—why some look beautiful and some not. These questions always come into my mind. Animals are very intelligent and superb engineers. I feel in certain cases animals are much more clever than humans. These courses changed my views greatly.

We learned that many animals sacrifice their lives to save others. For example, the male orb web spider makes the biggest sacrifice, giving his life for the health of his offspring.

I think the idea of life in Buddhism, Western science, and biology is very similar. Life is a progressive moment, a successive series of different moments; it moves from cause to cause, effect to effect, one point to another, one state to another, and everyone is connected to one another.

Self-sacrifice does not only mean suicide. In Buddhism it simply means to put others' needs before yours in an altruistic way. Giving your energy, time, and life for the greater good is sacrificing your life to serve the community.

Slime-mold cells searching each other out upon starvation look startlingly similar to human cells forming a new blood vessel in the early embryo, or to a human nerve cell searching for a brain connection, or to one of our immune-system cells on a mission to search out and destroy an infecting bacterium. These processes do not just *look* similar.

While people are obviously not slime molds, we are not as different as we might like to think. The processes and even some of the very molecules that slime molds use to move and develop from single cells to a complex multicellular organism are similar or identical to developmental processes in all multicellular organisms. In fact, the very same chemical signal, called cyclic adenosine monophosphate (CAMP), that slime molds use inside and between cells to attract each other and develop is also used as a major signal inside human cells.

Slime-mold cells responding to the starvation signal begin to line up behind the first cell, the closest and "hungriest" one that responded to the signal first. These cells move in a stream toward the original signal source.

Once the thousands of responding cells have gathered around, they make a developmental decision—a decision dependent on whether food is now available in the environment outside their cells. If it is, they form a many-celled slug that wanders off as a primitive multicellular searcher of food; if food is still lacking, they can further differentiate into a fruiting body full of spores, atop a stalk.

In either case the cells move around in a coordinated fashion, probably again determined by gradients of chemicals. In the slug, the cells are quite different than they were just a short time before. Now they have direct and coordinated relationships with each other; they orient themselves in relation to each other—some in the back, others in the front of the slug. They very directly *depend on each other*. Their differentness is an altruistic decision or commitment, in that each cell is no longer doing all that's required for life—as they recently did as single cells—but have given up some things to provide others. While their genes will live on via copies carried in other cells within the slug, they themselves may not.

If this mound of slime-mold cells does not sense the presence of any food, it develops into three distinct cell types—stalk cells, fruiting body cells, and spore cells. Which cells become which mostly depends on where they happened to be in their own cell cycles when they received the initial starvation signal, back when they were single cells. For example, most of the cells that later became stalk cells were in the stage of the cell cycle in which they were doubling their DNA to prepare for division.

The basic challenges of the slime-mold slug are similar to those of a human embryo: organize and situate thousands of genetically identical cells in space and time based on their past history to set the stage for a new and complex organism with diverse cell types.

The slime mold's story—one that has been occurring on forest floors worldwide since long before humans were around to walk by—is a classic representation of all that is developmental biology: life, death, and sacrifice. The basic story is the same for humans. The development of each life is a balance of living cells (some dividing, some not) and dying cells (some irreplaceable, some not).

The stalk cells sacrifice themselves for their genetically identical spore cells. Their genes live on. This is the ultimate sacrifice: cell death occurs so life can happen. Cell death occurs, as we shall see, throughout development. Such sacrifice is usually, as here, a response to the environment.

John Bonner made the video of the slime mold we watched in Dharamsala. He has studied these cells for over forty years. Many of the great scientists like John Bonner remind us of monks and nuns. They speak of their work with a kind of reverence; they live with and in their work; they strive to "become one" with what they study; they see beauty.

We monks and nuns especially admire Gregor Mendel, who was a scientist and also a monk. Mendel inspires and motivates us not only because

FIGURE 2.3 Mitochondria have external and internal membranes.

he is a monk like us and did such important experiments in genetics, but because he used very simple tools and logic to analyze his experimental data. Today, scientists use very technical methods and tools to study chromosomes and genes. The fact that Mendel, like John Bonner, didn't use such instruments and carried out his work in his monastery garden indicates that it is possible to study nature and make significant discoveries with careful observation and logic, without necessarily spending hours in a special laboratory full of expensive equipment.

You can do experiments anywhere. Mendel's work is also inspirational to us because he demonstrates that religious people can be scientists and vice versa. Mendel had infinite patience, growing and analyzing at least twenty-eight thousand pea plants over many years to reach his final conclusions about the fundamental laws of inheritance. Mendel's life story also motivates us. He came from a farming family, from simple beginnings. Many of us had similar humble beginnings—farms, nomadic families, remote Himalayan villages.

The very internal engines that keep both slime-mold and human cells alive—mitochondria—also are critical in cell death. The more our *cells* wear out and die, the more we as whole organisms age, and the closer we get to death.

Mitochondria (figure 2.3) live in our cells and produce the energy that keeps us going. Mitochondria most likely came about when "one day" in evolutionary time, one cell engulfed another cell, perhaps a bacterium, and then, instead of getting eaten, the engulfed cell stuck around, cranking out energy for its host cell.

We imagine this because mitochondria are themselves much like cells: they have their own membranes, their own DNA separate from the main, nuclear DNA of their host cell; and they divide on their own and have their own lives and population thriving within our cells. This is a neat little altruistic, symbiotic relationship. The mitochondria efficiently produce energy, while being protected and nurtured inside cells, and the cells have a convenient and effective energy source.

While you and your cells run on energy produced by mitochondria, side effects of this energy production include chemicals that can damage the very mitochondrial DNA and proteins that are required for producing that energy. These chemicals (reactive oxygen species, or ROS) have mitochondria both as their main *source* and their main *target*.

One theory of aging is, to simplify it a bit, that the longer mitochondria produce energy; that is, the longer we are alive, the more ROS they produce, the more damage to mitochondria they cause, and thus the more we age. We age, then, because our mitochondrial DNA and proteins accumulate more damage, become less efficient, and then pass on this decreased efficiency, resulting in less energy to "run" us.

Maybe it's not surprising then that mitochondria are also at the heart of cell death. One of the first things a cell does to kill itself is poke holes in its mitochondrial membranes. Out of the mitochondria pops one of the key proteins otherwise used and required for energy production and life promotion. But now this protein locates and activates a major death enzyme that in turn chews up cell parts, causing cell death.

So way down deep at the molecular level: the source of life is the source of death. Profoundly samsaric, no?

GIVEN ALL THE WORK they must do and all the energy they expend in attempting to reach an egg, sperm are of course chock full of mitochondria. We're back at the beginning of the chapter, back before a life begins. What can we learn about and apply to our own development from the processes illustrated by slime mold and the Samsaric principle of death as a crucial part of life? To find out, we follow our biological creation from before sperm and egg to the embryo to a developing young person.

One of the major causes of human infertility is dysfunctional sperm. Human sperm, even within one man, are different from each other, and the ones that have the most functional mitochondria are by any measure "the best" and most likely to fertilize the egg.[2]

Men continuously produce a lot of sperm. The processes from pre-sperm to mature sperm take about seventy days and involve basic cell division (mitosis) and also a special type of cell division called meiosis.

Meiosis results in two things: (1) the mixing of two sets of genomes (one from each of your parents) to increase diversity, and (2) sex cells —sperm or egg—that only have one (now mixed and unique) genome's worth of DNA. Eventually when one sperm and one egg meet, they re-establish a typical cell with two genomes that then divides to become most all of our other cells—all, again, with two genomes, one from each parent. Figure 2.4 shows meiosis for sperm (top), with representative chromosomes indicated, and meiosis for eggs (bottom).

As mentioned, two great waves of sperm death are programmed into the process of sperm development—one early to ensure the quality of sperm and the ratio of sperm to sperm-nurturing cells, and one after the sperm enter the reproductive tract of the female.

Fertilization seems to require a community effort of masses of sperm, virtually all of which sacrifice themselves in facilitating one getting the job done. All the sperm are tagged in such a way that, if they *are* attacked and destroyed by the female immune system, there is none of the inflammation that would normally occur during the typical immune-system response. Such inflammation would be dangerous for any potential fertilization and development to follow. Sperm capable of fertilization can remain in the female reproductive tract for up to a week.[3]

How does death ensure life in *mothers'* sex cells? Unlike fathers and their sperm, mothers are born with virtually all the eggs they will ever have (although some evidence suggests this is not entirely true).[4] Your mother is born with only a very small percentage of the thousands of eggs she started with when she was inside *her* mother.

Research in mice, another model system and one evolutionarily closer to humans than slime mold, gives us insight into how all this happens in females. As with sperm, not all eggs-to-be are created equal. Within the embryonic mother-to-be, a few dozen future eggs move in a discrete path—undoubtedly again following a chemical gradient— through the hindgut to a specific location. Along the way and also once they stop moving, these cells divide continually. In addition, they form bridges with other potential egg cells (as do developing sperm among themselves).

THE ENLIGHTENED GENE

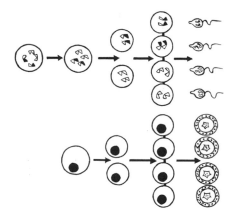

FIGURE 2.4 Meiosis, or production of germ cells—sperm on top, eggs below. For the sperm, chromosomes are indicated; the number of chromosomes is halved in the meiosis of all germ cells.

Eventually, these developing eggs stop dividing via mitosis and begin doing so via meiosis, the same process necessary for sperm production (figure 2.4). Again, meiosis mixes the DNA and divides in half the genetic complement all other human cells have. The potential eggs are called oocytes now; the bridges between them dissolve, and they are encapsulated in a follicle prior to birth. Death of some of these oocytes occurs throughout this process, but to the greatest extent at two specific times: right between when mitosis stops and meiosis starts, and before the oocytes are encapsulated in follicles.[5]

Human eggs complete the first half of meiosis one at a time every month, after puberty and until menopause, as one is released for potential fertilization. The second and last part of meiosis is completed only after fertilization.

Why so much death to make life? Evolution is about survival of the fittest, and this is true up and down the staircase—at the level of the cell, the organism, and the population. Both life and death are built into this principle. And this survival, this selection of healthy egg and sperm, is the beginning of life, certainly one of *the* most vital things to get right—selection of the potential next generation.[6]

In the case of egg production, many potential eggs probably die because of a lack of integrity—mistakes in the DNA sequence, or not enough cellular materials to support a future life. The egg must provide not only half a complement of high-integrity DNA, but also much of the other starting materials—RNA, protein, and energy—for the burst of life that directly follows fertilization. When these and other oocytes sacrifice themselves, they may do so in order to enable the removal of the bridges between oocytes, a step that is needed for egg development to occur, and to provide some of the additional necessary starting materials for their neighbor oocytes.[7] The many deaths of potential eggs is a kind of sacrifice, because it both allows sharing of these vital materials and, at the same time, prevents "bad" genes (that could result in faulty development and early death) from being passed on—thus increasing the chances of "good" genes and cells being passed on into a reproductively competent organism.

Only with substantial cause and cooperative conditions can something be produced. Neither an egg nor a sperm is a baby. A baby is formed when the sperm reaches the egg and fertilizes it, but this only produces tissues, organs, and body. According to the Buddhist perspective, it cannot be a baby until consciousness enters into it. Karma induces penetration of consciousness into a fertilized egg inside the womb.

Cell death enables you to see—again a vital part of development rarely focused on when taught and studied in the West. Apoptosis during development allows the necessary detachment of the developing eye lens from the surface it is attached to early in development. Apoptosis also results in the unique curved shape of your eye lenses.[8] Life, death, detachment, vision.

The Samsaric pattern of life, death, and sacrifice continues after fertilization, during the formation of an embryo, and in the next steps in development.

As the newly fertilized egg cell divides, and then the two resultant cells divide, and the process repeats over and over to make an embryo, apoptosis occurs *almost from the beginning* when some of the support cells that nurture the fertilized egg begin to die.

The developing embryo is a veritable trove of living and dying cells —the balance of proliferating, growing, and dying cells continues. Developing organs owe nearly as much to dying cells as they do to living cells. In nearly every organ that has been studied—often in other animals besides humans for ethical reasons—apoptosis is vital. Cells are pruned, winnowed, and sculpted all throughout development and sometimes into adulthood.

FIGURE 2.5 During hand development, cells of the "webbing" die, leaving separate digits.

Much of cell death in embryogenesis seems to have evolved to make life easier—more energy- and information-efficient. It turns out that it is much easier, requiring much less genetic information, to make an excess of what is needed and then kill off what does not work, fit, or connect, than it is to make just the right amount needed and keep everything that is made.

Imagine we want twenty-five nerve cells to connect to twenty-five specific different targets in the brain. Fairly simple mathematical models demonstrate that the most efficient way by far to do this is to produce many more than twenty-five nerve cells that are *not* specific, let them find their way to the targets, and then keep only those cells that find their way. The much more energy-and-information-costly alternative is to make exactly twenty-five specific neurons, each of which must find just the right target out of twenty-five options.

Following this general principle that death evolves alongside of and to allow life, many cases of apoptosis in development involve sculpting or elimination of excess. A famous case is the cell death that occurs in the "webbing" between developing fingers and toes, resulting in free and separate digits (figure 2.5). Apoptosis also sculpts developing limbs. Cell death shapes the four chambers of the heart[9] and sculpts the lung and skeletal muscle.[10] Long after you have grown up, apoptosis continues to help balance out the production of new skin cells emerging from cell division in order to keep the thickness of your skin just right and kill off cancerous cells.[11]

Immune systems have evolved over the eons to protect organisms from foreign invaders, that is, to prevent ill health or death. Ironically, as in the other processes discussed so far, the development of this death-defying system itself involves the sacrifice of many, many cells.

Making an immune system is a stunning example of producing more life than needed and then cutting back with carefully regulated death. How is it possible that immune systems specifically recognize foreign invaders such as bacteria and viruses *never seen before* by that organism? And how can immune systems then destroy only those foreign agents and not parts of themselves, their own organisms?

This process is so intriguing and complex that seven Nobel Prizes have been awarded so far for its elucidation. In a simplified version, each of us develops two interacting types of immune systems. One type, the innate immune system, recognizes features of pathogens that are shared by most of them. The innate immune system responds to foreign agents immediately and signals the other type of immune system, the adaptive system, with information concerning the specific nature of the invader. Now the adaptive system sorts through its vast preexisting set of potential weapons that can target this specific invader. These weapons are cells that happen to be able to specifically recognize and destroy this particular invader. Once such a weapon cell is identified among our vast stash, it is stimulated to divide and make many more identical copies of itself to go and help attack the invader.

How do we make this array of weapon cells in the first place? During the development of these cells, parts of their DNA are mixed and matched to encode thousands and thousands of unique proteins—one for each cell. The weapon cells are all more or less the same except for the unique protein that they carry, which might one day be called on to recognize and help destroy an invader. Once a specific weapon has been called on and used, an increased number of these weapon cells are kept around, and in this way our immune system "remembers" the infection and can more quickly respond to it next time.

As incredible and effective as the immune system is working this way, we are still left with at least one big problem: what if the unique immune-system proteins—the essential components of the specific search-and-destroy mechanism—happen to recognize *your own* cells? Then the immune system will attack and kill cells of its own organism. When this happens, autoimmune diseases—such as multiple sclerosis or rheumatoid arthritis—may result.

During development in the fetus and throughout life, as your weapon cells mature, they are exposed to the constituents of your own cells. If a match occurs, if an immune-system cell recognizes any part of another cell, development of that immune-system cell ceases and it kills itself. Because of such self-recognition, *as many as 95 percent* of the immune-system cells that compose this stash of weapons die during development. Sound familiar?

IN DHARAMSALA we were teaching the monastics about the immune system. They were having problems with our weapons, war, and foreign-invaders analogies—the immune-system analogies Western biologists and students take to naturally. Maybe it was even hindering their understanding of the immune system. Again, how we teach, think, and learn is profoundly worldview dependent; science is not just a set of facts, but a set of facts thought of, tested, discovered, and learned in a particular context. Now, whether teaching monks and nuns or American college students, we have another very different metaphorical and literal framework from which to work. Thinking about the immune system instead as part of the cycle of life, death, and sacrifice changes the conversation. Same facts, different framing. Some of our cells die because of infection or because they recognize self, sacrificing themselves so our other cells, and we as organisms, can continue to live. Later in the book, we investigate other striking environmental influences that "teach" and shape the development of the immune system and in turn teach and shape that system's partners in life, death, and sacrifice: the nervous system and the "good bacteria" that live in and with us.

A BALANCE OF life, death, and sacrifice is vital before, after, and throughout life. Again, the dynamic, cyclical nature of these processes—how vital death is for life in everyday functioning in processes as fundamental as learning itself—and its implications has only recently come to be appreciated.

We often learn a lot about what happens normally when we study the diseases that result from intentionally knocking the cycle and normal pattern out of balance or studying unfortunate cases where this has already happened in nature. Cancer is a classic example of such an imbalance; it can result from too much life (cell division) or not enough apoptosis (death) and sacrifice. Another intriguing example occurs in some cases of autism.

When I was born, about one in ten thousand children in the United States were diagnosed with autism; now it's one in ninety. What happened? Many different things, including better definitions and diagnosis, environmental changes, potential immune-system effects (to be discussed later), and increased understanding of autism. The most common genetically inherited disease on the autism spectrum is fragile X syndrome. Fragile X often leads to severe mental retardation. Why?

One clue comes from studies with fruit flies carrying the analogous gene mutation that in humans results in fragile X. Such flies have characteristics very similar to humans with the syndrome: complex behaviors are altered, nerve cells form aberrant connections, and immediate and short-term memory are lost.[12]

A specific set of neurons—that in normal individuals appears only early in development and then dies later—in mutant flies instead does not die, but is maintained throughout life. The balance of birth, growth, and death of neurons is knocked out of whack, with a severe result. Here the disease in a model system allows us to better understand what *normally* occurs and how in flies, and by analogy, perhaps in humans.

Excessive life, overproduction of cells, followed by pruning, scaling, cutting back, and death, allows for a more *efficient* version of life, but it also simultaneously allows for more possibilities, more adaptability, more flexibility, and thus the greater likelihood of survival. Recall our example of the twenty-five neurons targeting twenty-five specific targets.

Producing an excess of nerve cells (neurons) and then "letting" them all search for the targets also results in built-in flexibility and different possible outcomes *depending on the environment* those cells and that organism experience. Just as immune cells are eliminated if they experience self (and thus sacrifice themselves for the good of the organism), developing neurons—whether in the fetus or adult—are born, grow or not, connect or not to other neurons, and die or not depending on their environment. And as with immune cells, neurons are often born in excess and then pared back or, as in the case discussed previously, entirely eliminated, depending on environmental experience in time and space.

Think of the life-and-death balance of immune-system cells and its resulting changes and memory of infection as "learning" at the cellular level. Neurons are the stuff of our brains and nervous system, so their physical connections, their life and death, their learning and memories correspond, cause, and relate to our actual human learning. Learning and memory correspond to actual physical changes in neurons. If discoveries in flies are true for humans, someone with fragile X has *too many* neurons and neuronal connections.

When people learn, there is a corresponding fine-tuning of neuronal growth and connections, nerve cells growing and dying. When we think certain thoughts or respond in certain ways, the neurons in particular parts of our brains are activated or kept inactive. Memories must be somehow stored physically in our neurons. Neuroscientists understand these things at a general level and in some cases with some specificity, but are a long way from fully understanding learning at the cellular and molecular levels.

Not only building and having the *capacity* to learn, but also *learning itself* involves the development of new neurons and the death of old ones. This happens across developmental time, as is the case with fragile X, as well as in real time. Apparently, the same or similar processes are involved in both time frames. Neurons and their connections develop within a limited space-and-time framework, but with flexibility built in via overproduction. A balance of neuronal life, death, and connection integrates limitation and flexibility. An initial dose of life, an overproduction of cells and connections, followed by death and pruning back (all in response to experience and environment) optimizes efficiency and adaptive flexibility.

The story of the hippocampus provides a striking example. In the mammalian brain, the hippocampus, named from the Greek for its seahorse shape, physically changes in response to spatial learning. During development, as hippocampal cells are differentiating and dividing to make new cells, about half of these original newborn cells die. Those that survive mature. Later, in adults, these cells can again divide to make new hippocampal cells; their division is intimately connected to learning. The rate of division is related to the amount learned; and if new cell production is artificially blocked, learning decreases. Also, learning increases cell division and the survival of new neurons.[13]

The strange part is the connection to neuron death. Rats with the *fewest new* hippocampal cells have the best memory. Learning allows a greater number of mature neurons to survive better, causes less-mature neurons to die, and increases cell division of the survivors. If cell death is blocked, so is memory and learning; this neuron cell death is key at a particular time during the learning process.

Rats acquire spatial learning in two general phases: early in the task their improvement in performance occurs rapidly; then later, the rate of improvement levels off. It is during this latter phase of learning that cell death is vital. The cells that must die to allow learning and memory are newborn cells. So most likely, as in development, the cell death here involves "pruning" cells that are newborn, naïve, and unconnected, thus saving the mature, already-connected cells that represent the learned experience, now stored in the hippocampus.[14]

Thus, across both developmental and learning time, the physical substrates of learning—neural cells and their connections—divide, expand, and die depending on their maturity and on the environment.

IMAGINE IF the capacity for such learning itself, not just the genes involved in it, could also be passed down through the generations.

Brian Dias is a neuroscientist who teaches the monks and nuns in our project. Brian has a tall, commanding presence. His deep voice resonates with the accent of one born in Goa, a state on India's southern coast colonized by the Portuguese. Now a professor at Emory, Brian and his advisor at the time, Kerry Ressler, wove one of the most talked-about research stories of recent years.

Dias used the sense of smell in mice as a model to show that experience can alter neuron growth and death, *and* he uncovered one additional step—the sensitivity to the experience is passed on to the next generations. When mice repeatedly experience the pairing of a particular smell with an electrical shock, the smell neurons that grow in their brains are greatly increased in number—that is, the balance of living cells and their specificity are adjusted based on experience. Particular smells are known to associate and activate the gene expression and growth of specific olfactory sensory receptor neurons. In Brian's experiments, he used a chemical smell that is known to activate a specific neuron.[15]

In these experiments, the increased number of smell neurons linked to the particular smell used—connecting the *physical change* in cells to the *experience* of one smell and a shock—was then passed on to the off-spring of the initial animal. This means that males who were entrained to have a shock response in the presence of a particular smell had off-spring who also had a shock response to the same smell, *even though those offspring had never previously been exposed to that smell.*

How can this happen? An experience of a parent results in altered capacities of their offspring and of following generations that could not be because of gene changes. This is known as transgenerational epi-genetics, and we will elaborate on it and the controversy surrounding it more later. Brian showed that in the sperm of the initially entrained mice, the gene for the particular neuron receptor of the odor used, while identical in DNA sequence, is chemically modified (as compared to sperm from males who have not been so entrained) in ways that are predicted to alter its expression. So far, at least two generations of mice maintain sensitivity to this experience (that they themselves did not have) in their neurons and genes.[16] And we begin to see the intriguing, and troubling, implications of such work, that traumatic experiences —or at least sensitivity to them—can be remembered in the molecules and passed on to the next generations. We pick up on this trope in chapter 4.

PRESUMABLY, classroom learning too is reflected in dynamic cell activ-ity—although much of this learning appears to be more about the in-crease and decrease of *connections* of preexisting neurons, rather than necessarily the birth of entire new neurons, and is unlikely to be passed on across generations. Indeed, other research from the past decade shows that even in mundane, everyday experiences, brain cells change with and reflect those experiences.

We can now begin to weave together our knowledge in educational theory, psychology, and cognitive science with neuroscience such as Dias and Ressler's to understand and improve learning. Could such knowledge inform the challenges of teaching monks and nuns about science or of teaching scientists about Buddhism, or even give specific insights into individual learning? Probably.

The new and evolving field of educational neuroscience is taking the first tentative steps toward addressing this question by at first simply mixing the ingredients of different disciplinary approaches to learning.

At this point, researchers can at least thicken the soup by testing educational theories—previously argued primarily in the abstract or only by observing *behaviors*—in combination with observing *brains* in action.

Educational theorists have competing ideas, for example, of how learning two languages at once works, or of how people learn science concepts most effectively. Now neuroscientists can watch people's brains in action at the gross level, and since they know which parts of the brain are generally involved in certain functions, begin to make inferences. When one part of the brain "lights up" or becomes active, a known function of that part of the brain is most likely involved in whatever the person is doing. Such experiments are at this point very limited, but they do give an idea of the potential of this type of research.

HOW ABOUT WHEN a Tibetan monk, or anyone, learns something that deeply conflicts with his previously held ideas? Konchok recalls those monks back in his monastery to whom he can tell things a hundred times and, even in the face of concrete evidence, do not change their ideas. Nearly everyone has ideas they cling to with similar certainty. Does it make sense to first identify and make explicit the point of conflict before attempting to resolve it? What might be effective strategies for doing so?

The psychologist Drew Westen gained renown among politicians and scientists alike when he and his colleagues studied the brains of people presented with reasoning tasks that included information threatening to their favorite political candidate (as compared to the opposing candidate).[17] Such emotionally linked reasoning activates whole areas of the brain that are not activated when people are reasoning through ideas to which they have no emotional attachment. If an additional, different neurocircuitry is operating to deal with such conflicting information, perhaps different pedagogical strategies are required to identify, address, and reach and teach them. Now we can begin to develop and test new tools or old tools used in new ways, and watch our brains "learn" (or not) in response to different pedagogical approaches.

In similar experiments but in a different context, scientists were able to watch the change in activity (remember: change in activity corresponds to change in neuronal activity and perhaps also in neuronal growth, life, and death) in students who had mastered difficult concepts in the science classroom. Many scientific concepts are not intuitive—that a big ball and a small ball fall off the roof at the same rate (Galileo's famous experiment), for example—and when we look at students' brains as they observe such phenomena that do not fit their views, we can get a measurable sense of "what they're thinking" to give us ideas about how to improve learning, or at least to measure correlates to improved learning.[18]

When knowledgeable or naïve physics students watch a video of the big and small balls falling that *fits* their view of how things work, their brains have increased activity in their frontal cortex. The physics-naïve students only have increased activity in this area when the video shows the big ball (incorrectly) falling faster. So such increase in brain activity might be a "marker" for grasping concepts.

When either physics-knowledgeable or physics-naïve students see the phenomenon they think of as incorrect, parts of their brains that signal errors and conflict become active. Interestingly, the knowledgeable physics students have frontal cortex activation when viewing either the correct or incorrect video; but when viewing the incorrect one, they also have activity in the conflict and error-recognizing parts of the brain, suggesting their former beliefs are still there but are inhibited.

Does such research provide avenues for potential collaboration and enhancement in our project with the monks? Can we further illuminate such questions as: Does the cultural framework in which one learns alter not just *how*, but *what* one learns? Since the monastics have over centuries gained much expertise in training the brain, are there hints they can give us about such training to enhance or elaborate teaching and learning as we know it?

From the early age of seven, I left my home and family behind and went to India to become a monk. The transition from layperson to monk, and monk to science student, was not easy. While growing up, the questions that always bothered me were: Why did I become a monk? Who has made me a monk?

Another question: Is it my genes that made me a monk? This question came into my thoughts frequently while I was studying genes and genetics. I learned it is our genes and genetic makeup that help determine the way we act, and our genetics provide a foundation for our growth and development.

I mix science and Buddhism. I have come to realize it was not only my genes that made me a monk, rather it was also because of the environment, culture, and family. Sometimes parents and family put you on a path that you had never expected and it changes your life. Family can influence us in many ways. If our families are religious, we might take on those beliefs.

I grew up with people and in an environment strongly connected to religious faith and belief. Therefore, the majority of people in my village strongly believe in religion, monks, and priests. They consider it good fortune to have a monk in their village or family. Whenever people suffer or fall ill, the first option is to go see the religious person or monk. In fact, religious people play a big role in removing obstacles, healing, performing rites, and giving herbal medicine.

Now I have been a monk for twenty years. Honestly, I never thought of becoming a monk when I was very young; I wasn't ready. I was always busy with my childhood routine, playing with friends, collecting firewood, fetching water from a distant spring, looking after cattle.

Only those who have accumulated the karma in previous lives can become a monk. I am a monk in this life, and my Buddhism says the seeds of my becoming a monk were sown in my previous lives.

We have been focusing on "normal" death, sacrifice, and life in development—death that is actually a necessary part of life. But what happens when there is massive damage and cell death caused by external stress: physical wounding, say, or restriction of oxygen, such as that resulting from a stroke, leading to cell destruction? And are the realms of the mind and brain—such as the learning we just discussed, or thinking, meditating, and mental illness—possible to integrate with the realms of the physical body, so that one can positively affect the other and make possible or enrich healing of physical damage caused by external stress? We save this last question for chapter 7.

Regeneration is a special kind of rebirth. We discussed making, sculpting, and growing new cells, but how about re-creating whole new tissues and organs that have been destroyed? The hydra can regrow an entire head; newts, their tails. The human liver can fully recover and "grow back" even when 75 percent of its cells have been killed (most commonly by alcohol or hepatitis). And our skin, intestines, kidneys, muscles, and nervous systems have varying degrees of regenerative capacity. Why can only particular tissues regenerate and only to a particular extent? If we could replace dead or damaged spinal tissue, for example, perhaps paralyzed people could walk again.

Stem cells are key players in regeneration. Stem cells from embryos have the capacity to become any kind of cell (thus, the political and ethical controversy over obtaining human embryonic stem cells from aborted embryos or from embryos made but never implanted and currently in freezers of in vitro fertilization clinics), but adults have many stem cells as well. These adult cells produce a more limited diversity of cell types. Think of these stem cells as "reserve cells," stored for regular, routine replenishment or emergency replacement of other specific cells after damage.

The smell or olfactory system is a striking example of "routine replenishment" of stem cells. In mammals, thousands of smell neurons (such as the ones described previously that were studied in conjunction with shock in mice) combine to allow the detection of many odors. These neurons are continually replaced throughout life by cell division of olfactory stem cells. These stem cells can give birth to many diverse olfactory sensory neurons. Each new cell is born and differentiates with the specific information to be able to bind only particular odors and to make the proper and exact connections within the olfactory system.

Stem cells have a kind of dynamic, Buddhist-like nature about them. Depending on environmental signals and the needs of an organism, stem cells can divide in three different ways: either by the usual symmetrical mitosis resulting in two new stem cells that are clones of the mother, or into two differentiated cell clones that are no longer stem cells, or asymmetrically into one cell that maintains the mother cell's stem cell identity and one that is differentiated.

The signals that tell stem cells to asymmetrically or symmetrically divide to produce new replacement cells often are released from dead and dying cells. In other words, *the dying cells themselves directly induce new life.* Dead and dying cells also release signal molecules that activate already-differentiated neighbor cells of the same type to divide and generate more cells of the same kind as those killed.[19]

From our growing knowledge of stem cells and development has emerged the field of regenerative medicine. The paths of this field diverge in the woods along the very same philosophical directions with which we began this chapter. One heads off in the direction of linear thinking, of technologies set on prolonging life and slowing aging as long as possible. Can we regenerate whole animals, even human beings, from one cell? Can we put off biological death forever? The other path circles back and asks: can we relieve the suffering of those already alive who have debilitating neurodegenerative diseases such as Alzheimer's or Parkinson's? The two paths are not mutually exclusive, of course; and as we walk both paths, beneath our feet and unbeknownst to most, roam the slime molds, tiny creatures from which we have learned much of what allows us to walk these paths in the first place, to understand how we become.

CHAPTER THREE

How Did Life Begin?

The other monks asked me, "What did the first cell come from?
What causes a cause?" And I had no answer at that time.
KONCHOK

Nothing in biology makes sense except in the light of evolution.
THEODOSIUS DOBZHANSKY

magine three dozen monks and nuns, heads shaven, clad in maroon and gold, scampering up and down mountainsides in the late-afternoon summer light, collecting soil, water, plant, and animal samples and chatting happily among themselves.

Giddiness is not the first emotion you probably associate with monks and nuns. But here we are in our first few days together nearly a decade ago, American scientists and Tibetan monastics, and giddiness is certainly in the air. Konchok and I have only just met. He stands out with his ready laugh and the distinctive gash of blue, representing his Bön sect, amid his maroon robes. He and the other monastics run up and down the hills, pulling up roots they recall using for medicinal teas in their childhood, hugging each other affectionately, grinning from ear to ear, and periodically grabbing a translator to explain to us what they have found or to ask us a question.

We are at the tree line just above Dharamsala, carrying out an experiment the monks and nuns designed based on our initial discussion about evolution. As the days go on, it becomes clearer that their happiness in the hills is about much more than being active scientists. Not only is it a rare opportunity for the monks and nuns to be outside and active beyond the boundaries of their monasteries and convents,

69

but it turns out many of them have not been in these Himalayas since they were young and living in Tibet. As children, they escaped over these mountains, eventually heading for monastic institutions far away in the south of India.

Who would have thought their first return to these peaks would be to perform an experiment they themselves had designed to test a foundational hypothesis of that most famous of Western scientists, Charles Darwin, a man many of these monks and nuns had never heard of until a few days previous?

That first summer we indeed began with evolution, Charles Darwin's brilliant and synthetic theory that underlies all of biology. What better way to introduce life's definition, origin, and basic molecules; the central concepts of time and relatedness; and the eternal conversation between living organisms and their environments than through a discussion of the idea that ties them all together?

The irony doesn't escape us. Here we are teaching evolution—a favorite bogeyman of many religious people back in America—to men and women whose whole life *is* religion.

The monks and nuns are at the tree line as part of an experiment they designed to test some predictions of evolutionary theory. Darwin's famous idea of natural selection posits that the environment has an impact on organisms, that the environment "selects" organisms with the traits which allow them to be successful. In the evolutionary sense, "success" means living long enough to have offspring—the more offspring, the more successful the organism. Despite the terms "selection" and "success," in the Darwinian view, natural selection is not purposeful (nor is it random); nature selects for whichever traits organisms have that *happen to give them an advantage.* This is in contrast to the (in)famous ideas of Jean-Baptiste Lamarck, who proposed that individual organisms change their traits *in response to* the environment and then pass those changes on to their offspring; a giraffe stretches and stretches her neck, making it longer and then this length is passed on to her offspring, while organs she doesn't use shrink and their new shrunken size is also passed on.

That night Dawa, one of my classmates, and I had hot discussions on Darwin and Lamarck. I really preferred Darwin's ideas. Dawa was against

Darwin and believed Lamarck, that characteristics acquired by an organism during its own lifetime would be inherited by the next generation. He said that this is similar to what Buddhists believe: during our present lifetime, if we do something negative or positive, the result will be passed on or appear in our next life. But I said Larmarck was not right. For instance, if in your present life you paint your hair red, you will not pass this red on to your offspring or to your next life. I also made a clarification to Dawa about Lamarck's ideas: acquired characteristics, according to Lamarck, are passed to the next generation, not the next life. Later we learned (see Dias's experiments in chapter 3 and also the discussion in chapter 4) about new discoveries in a field called epigenetics that say maybe the answers lie somewhere between Darwin and Lamarck.

We introduce the monastics to the scientific approach to knowledge: identify the problem or question, develop a hypothesis to explain it, design and perform an experiment to test the hypothesis, draw conclusions, and repeat. If the environment affects organisms, the monastics reason, then different environments should be home to different organisms or to similar ones with different traits.

We pile the monks and nuns into vans and take them to three extremely different elevations, from lower to higher. As we wind up Dharamsala's interminably twisting and narrow roads, I try not to lose my lunch, while the monastics divide into groups: one to collect and observe insects at each location, one to do the same with plants, one to observe soil, another air, another water. As the air thins, the energy among the monastics stays thick and positive—a buzz of conversation and questions. Nonstop questions. The translators can't keep up: Why is this bug this color? Is this moss a plant or an animal? What type of tree is this? Why are there no trees here? If plants are sentient, why can't they move?

During our experiment in the mountains, we were so happy, running, jumping, and collecting plants, roots, and dead insects. We played a lot with nature.

When we went outside to do those experiments testing Darwin's ideas about organisms and the environment, this was the first time we monks had seen adaptations in nature, our first experience seeing this variation in different environments.

After we did that, most of the monks felt more encouragement, more enthusiasm. In the lecture, we learned about variation, inheriting traits, but outside, what we saw, we realized this was real.

Not only did we enjoy the environment a lot, but the main thing was going outside and looking at the environment, and looking at different species—how they were different at different elevations—and looking at these things in a "scientist's way." Looking like that, we had the feeling *we* could be scientists examining the natural world with respect and not disrupting or destroying nature.

Upon returning to the classroom, the groups synthesize, write up, present, and excitedly discuss their results. They sit cross-legged on the floor in tight circles—the nuns, still not used to engaging closely with the monks, shy and hanging back, but attentive—with samples and discoveries strewn before them. Their findings are consistent with environment-organism interaction. Plants at higher elevation have different traits from those at lower elevations; generally, they are shorter or smaller. And the entire absence of trees is evident, of course, above the tree line. Ants in moist or dry, high- or low-elevation environments have traits that appear to have adapted to fit those environments. Adaptation indeed is the result of natural selection.

IN THE UNITED STATES and other countries, scientists' willingness to engage or even mention religion has become increasingly rare. Science teachers from high school to college—to avoid controversy, or citing lack of expertise outside of science, or for fear of losing their jobs—not only have stopped mentioning connections between religion and science in the classroom, but some have stopped teaching evolution altogether. This is a disaster for an idea at the heart of all biology and medicine.

Darwin himself waited a long time before publishing his theories, and one reason was his fear of how the church, of which his wife was a close adherent, would respond. But the current severe degree of tension between science and religion is more recent. In the early 1970s fruit-fly biologist Theodosius Dobzhansky made a strong argument for the consonance of science and religion in a talk for the National Association of Biology Teachers. Ironically, scientists love to cite Dobzhansky's title of the talk, which leads this chapter: "Nothing makes sense except in the light of evolution"—without, I'm sure, realizing what the talk was all about, and that Dobzhansky goes on later in this talk to say, "I am a creationist *and* an evolutionist. Evolution is God's, or Nature's, method of creation."[1]

OUR PROJECT with the Tibetan monastics provides avenues into teaching and learning evolution in relation to creationism, and science and religion issues in general, in a Judeo-Christian world. This is a powerful example from our project of the ping-pong insight from chapter 1, the one about tensions as entry points for discovery. Rather than *avoiding* points of tension, let's first *identify* them and then use them to enrich discussion and learning. Listen to philosopher Pierre Teilhard de Chardin, a Christian, whom Dobzhansky quotes in that talk he gave to the biology teachers: "Evolution is a light which illuminates *all* facts, a trajectory which *all* lines of thought must follow [emphases added]."[2] So it makes absolutely no sense to leave out evolution altogether, much less not to engage it fully head-on.

HOW DID life begin?

Even my mother-in-law, who is wary of all things science and is so far right of me as to hardly be visible, is cool with natural selection. If I don't call it that and say, "The squirrels that best remember where their nuts are buried will outlive the ones that don't and will therefore pass their traits on more often," she gets it. But where it starts to get sticky for many—even scientists—is when you ask questions such as, "How did life begin?" How does Darwin explain that?

WE WERE TALKING with the Dalai Lama at his home in Dharamsala.

A giant golden Buddha, with peaceful eyes, looked on. I had just noticed a hole a moth had eaten in my seldom-worn suit jacket; I tried to feel compassion for him—the moth—given the circumstances. Minions hovered—translators, security, people with a lot of phones looking serious. But the Dalai Lama's eyes were dancing. He sat in his usual cloud of calm and his sensible shoes, smiling. He is wisdom and playfulness at once.

I asked him about risk. Wasn't he taking a chance bringing science into the monastic curriculum, shaking things up after half a dozen centuries? No risk, he said, "Buddha himself made it very clear that his teaching should not be accepted out of faith, out of devotion, but rather out of thorough investigation and experiment. So one casualty is Mt. Meru. Not a serious matter [laughter]!"

Traditional Buddhism (as well as Hinduism and Jainism) teaches that Mt. Meru sits at the center of the universe, and all creation springs from it. The Dalai Lama had just laughed aside an entire mountain of mythic proportions. He does not dismiss the Buddha's teachings, of course, but says the purpose of the Buddha is to liberate sentient beings from suffering, and by implication, not to explain, or even perhaps worry about, the intricacies of creation. The Dalai Lama often says that if science can logically demonstrate a concept counter to what he previously had thought to be true, he will change what he believes. He focuses on the spiritual message of Buddhism and leaves room for the accommodation of new knowledge. The Dalai Lama was telling me in his lighthearted, serious way: Mt. Meru is a story; I accept evolution.

Konchok leads this chapter with perhaps the biggest potential problem, the most sensitive tension point, in teaching evolution—the origin of life: *What did the first cell come from?* Can the principles of evolution explain how life began? How about that first cell? This is a challenge, of course, not just for Buddhist monks and nuns, but for people of all religions, and for many scientists—including Darwin. Dobzhansky did not directly address life's origins in his 1973 talk. Even if life can be created from scratch in a test tube, we will never be sure how it actually began. But we can look for clues in the geologic record, in laboratory experiments, and in currently living organisms and environments.

In the Dharamsala classroom with the monastics, we discuss early explanations of the origin of life, such as spontaneous generation. It was only 150 years ago that Louis Pasteur and Lazzaro Spallanzani finally disproved this idea that life could start at any time—say, in the kitchen or the lab—on its own without any precursors. We tell the monastics about the famous experiment shown here. Bacterial media is boiled in two curved-necked flasks that let in air but not microbes. For the first flask (A in figure 3.1), nothing further is done. For the other (B in figure 3.1), its neck is broken. Bacteria soon grow in the broken-necked flask, but even a year later no bacteria have grown in the media of the flask with its neck intact.

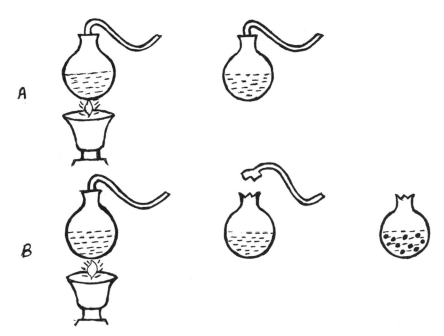

FIGURE 3.1 Spontaneous generation disproved. In A and B, water is boiled to kill any living material; then only B is opened to the air; then, wait, and living matter grows only in B.

In the front row, Kalsang's hand shoots up. Kalsang is one of Konchok's friends and classmates. From the beginning, it is clear that Kalsang is a natural-born scientist. He listens and follows closely. Early every morning before classes, Kalsang's multiphonic chanting leads prayers, his voice resonating through the temple in ways that surely transform both his and his listeners' physiology. Later we learn that Kalsang is also a renowned scholar of Tibetan history.

Now, with the twinkle in his eye he often has in class, Kalsang tells a story through the translator. Once he boiled water and added it to tea leaves in a thermos and then screwed the lid on tight. Then he lost the thermos, finding it again only weeks later. When he opened the thermos, it was full of noxious-smelling microbes. Wasn't this at odds with Pasteur's flask experiment? Microbes must have spontaneously generated in Kalsang's thermos.

Kalsang raises subtle but important points. To entirely exclude microbial growth, one must (a) boil all constituents *for long enough to ensure all microbes die*, and (b) after boiling, the material cannot be re-exposed to air, since air is full of spores and other "seeds" of microbial life. More than likely, in Kalsang's "experiment" neither of these conditions had been satisfied. His question demonstrates the kind of keen logic and skepticism many of the monks and nuns bring to the study of science.

The human mind desperately seeks solutions to big mysteries and hangs on to old favorites tenaciously. Years after Kalsang's question, teaching evolution and life's origins in the monasteries in south India, another hand shoots up: Did I know that leather turns into frogs? Scraps of leather left in a drawer in Tibet over the winter change into frogs when the weather warms. Half the monks in the class laugh nervously, others nod their heads in agreement.

The monks designed an experiment. They put pieces of leather in a plastic bottle, holes poked in it for air just in case, and place it in a drawer. They have been checking the bottle for frogs since last year.

A few days into our teaching, the monsoons blow in. Suddenly frogs appear from nowhere, hopping across the fields surrounding the monastery, when the day before there were none. And the story becomes clearer.

Some species of frogs (and other animals) can bury themselves in the mud and simply shut down until conditions improve. Surely, across the eons, in some frigid Tibetan winter and not knowing a frog or two were in the soil, someone had tripped over some dirt or thrown a clod against a wall, or had some leather and dirt stored away, and upon warmer conditions, frogs had appeared as if from nowhere.

MANY INTRIGUING THEORIES, using really no more than Darwin's big idea of natural selection, posit how life may have started. A favorite of mine is from chemist and molecular biologist Graham Cairns-Smith; it is logical and has a kind of poetic flair to it.

Cairns-Smith begins with the analogy of the giant stone arches built by ancient humans long before modern heavy machinery existed. How could such imposing arches be built—the keystones put in place balancing the two massively heavy curving sides of it—by prehistoric peoples? The answer—which my students rarely, by the way, come to on their own without some hints—is to first create a pile of dirt the height

FIGURE 3.2 Cairns-Smith's analogy of how to build an arch from enormous stones in ancient times—an inspiring thought about how life may have begun on earth.

of the arch you want, then place stones in the desired shape and location to build the arch so that all the weight is borne by the dirt, and finally, once all the stones are in place, including the keystone, remove the dirt and there it is: a large, balanced arch (figure 3.2).[3]

In Cairns-Smith's origin-of-life model, the "dirt" is clay, which probably existed in those very early prelife days. Even though they are not living, clays have many characteristics, roughly speaking, of life. Clays form crystals, so they have a repeating structure that can "reproduce" (and they even do so spontaneously upon saturation in water, just like sugar crystals in the production of rock candy). Clay crystals can also "mutate" if new minerals or other components are added to them, changing their properties. Clays are also silicates, so they have sticky surfaces that tend to attract other chemical components.

So clays have a set of characteristics—stickiness to affect the flow of water or bind minerals, repeatability of structure, capacity to be windblown to new locations when dry—which can vary depending on their environment, on the particular type of clay, and on what is piggybacking on the clay. Different combinations and degrees of these characteristics are selected for by the environmental conditions. In millions of independent events over millennia, some of the materials piggyback-ing on the clay begin to collaborate and interact. After helping give particular clays an advantage for eons, at a certain point the piggyback-ers no longer need the clay—the dirt is removed, leaving the hard-to-believe arch.

The arch in this case would be a relatively simple "life molecule" such as RNA. From modern-day living examples, RNA alone is known to do pretty much everything life does—reproduce, mutate, adapt, use energy, catalyze reactions.

From simple experiments, we know of other very different types of molecules that spontaneously form enclosed "cells." Some of these primitive RNAs might have become protected by such a cell, and we go stepwise from there over millions and millions of years to primi-tive life.

Cairns-Smith devotes a whole book to this theory, so we have cer-tainly not done it justice here.[4] Again, the point is not that this is nec-essarily what happened, but that given Darwin's principles, and the amount of time and the environment known to exist based on geologic evidence, this or something like this *could reasonably* have occurred. As with the other evolutionary explanations discussed, this theory of the origin of life does not exclude the involvement of a divine hand, which may be part of Dobzhansky's point in his famous speech to sci-ence educators.

Why do turtles have bony or leathery shells on their backs? I think about this kind of question a lot.

There are many different explanations—scientific, religious, abstract, dogmatic. Some are based on evolutionary theory, some on cause-and-effect (karma) theory, and some on other religious perspectives. Evolution talks about how structure and function are very related, and in this context, the function of the solid leathery shell is to protect the turtle from preda-tors, improving his survival and, therefore, showing natural selection.

But karma theory explains that it's because of the turtle's previous life. In a previous life, the turtle had a house, but he never rented the rooms or let anyone share his home. That previous karma resulted in the punish-ment of his carrying around his home as a turtle for life. Even as I hear this, it doesn't really make sense, and questions come to my mind: Did all turtles own homes in their previous lives? Or did one turtle own a home in a previous life, and in this life he or she is manifested in many turtle forms? Why don't individual turtles accumulate different karma, or do they?

We all watch. But we do not observe. Humans have the potential to analyze what is fact and what is not. But usually, we do not really observe or truly understand. Unless we do, we won't come up with new ideas and innovations. Sometimes dogma does not allow us to accept truths.

We monks are taught: "Whether fully learned or not, each person has his or her own ultimate truth."

One monk asked me, "The goal of studying Buddhism is to gain enlightenment individually and along with other sentient beings, but what is the goal of studying Western science?"

As I discuss throughout this book, I see tons of basic similarities and dissimilarities between Tibetan Buddhism and modern science; but for me, as a Tibetan monastic studying Buddhist philosophy, the super important thing is to find and particularly investigate the most profound fundamental controversies between them in order to come up with new thoughts. I hope the scientists feel the same.

Are humans distinct from other organisms? Did we evolve from them? These are other sticky questions for many.

Tibetans are striking examples of natural selection.

I look out at the rows of monks and nuns, cross-legged, maroon-robed, thoughtful. There's a kind of communal afterglow in the room following their elevation-altering experiments and resultant presentations. In some eyes, I sense a new engagement: maybe there's something to all this . . . at least we got to go outside!

I balance a glass-bottled Pepsi in my hand, puzzled by the English inscription below the ubiquitous cursive brand name: "Contains no fruit." For days since my trek from the States, soft drinks have been all I can keep down.

I stand up: "You are the fastest-evolving humans on record!" The translator translates. The room quiets.

It's not easy to live at fourteen thousand feet and above. The greater the altitude, the less oxygen available, and humans (as well as all other aerobic organisms) require oxygen to produce energy and survive. Up to 40 percent less oxygen is in the air on the Tibetan plateau as compared to sea level. Tibetans have, relatively quickly in evolutionary terms, evolved a number of remarkable high-altitude adaptations.

When lowlanders travel up to high elevations, they make more red blood cells and hemoglobin, the cellular and molecular carriers of oxygen. But the cost of this is severe altitude sickness—fatigue and headaches—and low-birth-weight babies with poor health.

Tibetans adapted to avoid these problems *without* producing more red blood cells and hemoglobin and the associated problematic side effects, and they did so in less than three thousand years, fewer than one hundred generations.[5]

This all happened when the future Tibetans split from the future Han Chinese, the former heading for the Tibetan Plateau, and the latter for what is now mainland China. Only Tibetans with these adaptations survived, living longer and healthier, and thus having more children, than those without these traits. In other words, only the few members of the population who originally moved to the Tibetan Plateau, and happened to have traits that allowed them to live and thrive at high altitudes, survived. Existing traits (or ones that emerged because of the occurrence of new mutations or alterations of gene expression like those described in chapter 4) in the population were selected for and quickly became predominant.

Lowland Han Chinese have flooded into Tibet in recent years. The Chinese women have discovered they cannot give birth at such a high elevation (8,000 to 17,000 feet) without severe side effects. Babies born to two Chinese parents at high elevation weigh an average of 66 percent less than babies born to Tibetans at similar altitudes; preeclampsia is much more common in Chinese mothers giving birth at high elevation, and pre- and postnatal mortality in mothers and newborns is also significantly greater in Chinese than in Tibetans. So pregnant Chinese mothers now return to a lower elevation before having their children.[6]

Evolutionary biologists refer to this type of evolution as *microevolution*. Microevolution is when changes *within* populations allow adaptation to the environment. The argument for microevolution is less controversial than the *macro*evolution discussed shortly.

Note again that selection can act on only preexisting traits—in this case those that allow the Tibetans to thrive at high altitude. Also, while natural selection has no intentionality, the process is not random; it "selects" those existing traits that allow better survival. There is a "conversation" between the environment and the organisms, between the Tibetans' traits, those traits' capacities, and the diverse environment of the Tibetan Plateau. The Tibetans adapt to the high altitude; they domesticate native plants and animals; and all of these organisms, in turn, change the environment, which the humans and others then adapt to, and so on (the unique ecosystem of the Tibetan Plateau and its important implications are explored in chapter 5). This environment-organism conversation of natural selection only stops when a species goes extinct.

Francisco Ayala, renowned geneticist and former priest, calls this and others of Darwin's insights the second half of the Scientific Revolution that began with Copernicus's insight that our planet is not at the center of the universe. Darwin demonstrates that our species does not operate under different biological rules, nor are we at the center of or above all other organisms. Both scientists bring about a fundamental shift: the universe and life on earth can be explained by natural laws, systems that "human reason can explain without recourse to supernatural agencies."[7]

Since at least Aristotle, humans have been classifying organisms based on similarities—where they live, whether they have backbones or feathers. Darwin realized that, based on similarities and differences, one could build family trees which connect different species to common ancestors (his thinking is captured in a famous image of such a tree in his notebooks),[8] that one megafamily tree can connect all species to one common ancestor. In evolutionary terms, each organism and each species is a collection of traits selected by the environment over many generations; at different points, new species emerge because of environmental distance—separation in time and space. Let's pull these ideas apart.

We build the case *against* the idea that humans are different and distinct from all other organisms in three steps: first, all humans are related to each other; then, humans are related to monkeys; and finally, *all* organisms are related to each other. This universal relatedness, together with the concepts of adaptation and natural selection, compose Darwin's core principles of evolution.

AS RECENTLY AS 150 years ago, Louis Agassiz, one of Harvard's premier scientists at the time and a contemporary of Darwin, could argue that species are fixed and do not evolve into one another. He claimed that even humans are divided into distinct races, each of which emerged independently in a separate geographical location. This allowed him to develop some pretty racist ideas, including that the book of Genesis explains the creation of only the white race.[9]

Adrian Desmond and James Moore make a convincing case in *Darwin's Sacred Cause* that a major driver of Darwin's development of the theory of evolution was his abhorrence of slavery and the many cases of abuse of dark-skinned people he witnessed during his travels. The Darwin family had a long and active history in the abolitionist movement. If Charles Darwin could demonstrate that humans are all of the same species and that, contrary to the ideas of Agassiz and others, we all share a common ancestor, our shared brotherhood would strengthen the case against slavery and other abusive practices.[10]

A century and a half later, we know that humans, *Homo sapiens*, are all members of the same species. A species is a group of related organisms that are similar and can reproduce with each other but not with other species. And therefore, in the twenty-first century, very few people take issue with the fact that all humans share a common *human* ancestor. Nor do many disagree with the idea, generally speaking, of natural selection—a change in traits over time in relation to the environment—*within* one species. Most are even comfortable up to the point where such change might result in two closely related species. For example, the ants that the monks and nuns observe at very different elevations and environments are similar, but with traits that fit their particular environment. This makes intuitive sense.

Closer examination of how organisms and populations adapt and change helps build the argument for *universal* relatedness.

DARWIN AND many before him logically linked relatedness to similarity and difference. That is, the more similar, the more closely related.

A family tree of the monks and nuns consists of them linked to their parents, grandparents, and great-grandparents, all the way back to those first high-altitude-adapted Tibetans, then back to the common ancestors of both the Tibetans and the Han Chinese. Like each of us, the monks and nuns are the biological sum of all their ancestors, selected over generations. Far enough back in time, Tibetans share a common ancestor with all humans.

Some of the most powerful evidence for evolution has emerged over the centuries from the science of relatedness—from studying similarities in fossils, structures, functions, and lifestyles. Now we can see and study relatedness at the most basic levels: the cellular and molecular architecture that underlies all species. All of life and all the traits of life on earth are composed of and originate from the same basic molecules: nucleic acids (RNA and DNA), fats, carbohydrates, and proteins. Cells are the basic units of all life—from bacteria to mammals. We can now follow the relatedness—the relatedness that Darwin saw and intuited —down at the first level of the Living Staircase: the DNA that helps shape organisms' traits.

KONCHOK SPEAKS EARNESTLY of his and his fellow monastics' admiration for Gregor Mendel, one of the most famous monks who ever lived, and clearly someone with whom the monastics relate well. In his small garden and meticulous notebooks, Mendel provided experimental evidence with pea plants for the universal molecules of inheritance—what came to be called genes. Darwin and Mendel were contemporaries. Apparently, Darwin owned a copy of Mendel's work describing genes; but Darwin could not read German, so he never opened Mendel's manuscript. In fact, Mendel's research was not rediscovered until decades later.[11] Mendel demonstrated a basis for inheritance, something concrete on which Darwin's concept of natural selection could act.

Today, the picture is clearer. All genes in virtually all organisms are made of DNA. All DNA is composed of the same four chemicals in different order; these chemicals carry a code, which is read to make RNA. Then some of this RNA carries out various jobs directly, while other RNAs are read in triplets, each encoding one of twenty different amino acids. These twenty amino acids are linearly connected, in the order dictated in the DNA of genes and the RNA it encodes, to make proteins. Proteins in all living things are made of these same twenty amino acids in different amounts and order; these proteins fold into

complex three-dimensional structures that, together with fats and carbohydrates, account for the structures and functions of all cells and
thus of the organisms they compose.

The total DNA sequence—the genome—of every human is extremely similar to that of every other human—the genomes of Tibetans and Han Chinese especially so. Indeed, because the approximate
rate at which mutations occur in genes is known, geneticists are able
to determine that Tibetans and Han Chinese descended from common
ancestors a relatively recent three thousand years ago.

Since genes help determine traits, when *traits* were selected for
among the Tibetans, particular versions of *genes* for those traits were
likely selected. Which genes? Because of the side effects of giving birth
at high altitude, we would expect to see differences in genes involved
in processes such as fetal and maternal access to oxygen and maternal
blood pressure. When the genomes of modern-day Tibetans and Han
are compared, about thirty genes pop out that are strikingly different
between the two groups. One gene, for example, encodes a protein that
is turned on in *all* people under low-oxygen conditions. This protein
is involved in the production of red blood cells. Nearly 90 percent of
Tibetans have the same particular version (allele) of this gene while
barely 10 percent of Han do.[12]

Recall that while it seems "logical" to increase the number of red
blood cells at high altitudes, the side effects are severe, and thus natural selection "looks for" other ways to solve the low-oxygen problem.
Now that we know there is a particular allele involved in making red
blood cells that is predominant in modern Tibetans (and not in their
close relatives the Han Chinese), the hypothesis is that this allele was
selected for at high altitudes. Perhaps the protein this allele encodes
somehow *constrains* the production of red blood cells in Tibetans or
results in the production of red blood cells that are somehow different
from those of us lowlanders. If such an allele was also found in other
high-altitude populations, such as those in the Andes, this would be
more strong evidence for its role. Perhaps such research will be under-
taken by some young monk-scientist in the near future.

ONE SUMMER, after the sciences had become an official part of the monastic curriculum, we are teaching evolution to a group of one hundred

monks in south India. This group has already studied science for a few months and, prior to our coming, has been reading the evolution textbook we have written and translated. After a few days of extensive discussions, they seem to grasp Darwin's basic ideas. So we decide to get them working and thinking on their own, to have them ask their own questions, do some background Internet research, and present their findings to the class. We divide the monks up into groups of three or four with a mix of different degrees of knowledge in science and in English, and we ask each group to develop a question at the intersection of Buddhism and evolution. We meet with the groups individually to help them sharpen their questions, and then send them to the computer lab to do research with Sherub on the Internet.

Sherub is one of the monks who, like Konchok, spent several years at Emory. He is by far the quietest and shyest of the group. During spring or winter breaks in Atlanta, while the other monks traveled in America and spent time at the homes of new friends, Sherub stayed in his apartment, meticulously transcribing into English all the lectures he had tape-recorded from his Emory classes, translating them into Tibetan and studying them endlessly. It turns out Sherub is an accomplished poet and novelist, with some of his work actually in the Emory library. Sherub went on to teach monks science back at his monastery and established a science center there.

Somehow, with (1) the skills of a Tibetan instructional technologist discovered in Dharamsala, (2) a Rube Goldberg contraption of giant car batteries rigged up to prevent our twenty computers and our Internet service from going down when the power inevitably went out, and (3) a lot of luck, we have set up a computer lab for the monks at this monastery. And this is where the groups go to study and research their evolution and Buddhism questions with Sherub.

One group, led by a monastic I have dubbed (fondly) "Hollywood monk" in my mind, as he's the spitting image of a young Marlon Brando, has formulated a question based on a Tibetan myth. The story is that the original humans were giants and that over time they "evolved" into smaller and smaller humans, until they became the size we are now. Sounds a bit strange; however, many cultures have such "giant stories," and there is some resonance with Darwin's ideas if you imagine that in "the old days" the environment was rough and brutal. It might make sense that bigger, stronger humans would survive better in such environs. So off went Hollywood monk and his buddies, who had never before used a computer, much less the Internet, to explore this question.

The next day, the group reported back excitedly. They had found a website that perfectly supported their idea about ancient human giants! Everything and anything is available on the Internet, of course —even Photoshopped images of giant human skeletons dwarfing their modern-sized human "discoverers." Here was a teachable moment.

We discussed the manipulable nature of images and of data in general; we explored the nature of the evolution of scientific *knowledge* and how (ideally) it is sifted and challenged through peer review and replicability. Back they went to the Internet, this time with some better hints for identifying reliable resources. All of this—the nature of science, their first incorrect findings, and the stronger data they found later—went into the presentation of Hollywood monk's group to the class. They discussed their original hypothesis and the reasons for it, how they found the website describing human giants and our skepticism of that site, the power and danger of the Internet, and their conclusion that, despite the rumored existence of exceptionally large weapons and saddles, they had as yet found no credible evidence that humans had once been huge and had then evolved to become smaller over time.

NOW, AFTER GRASPING natural selection, adaptation, and the basic concepts of evolutionary relatedness, comes the giant leap, so to speak —one of the most stunning of Darwin's insights: all humans are related to each other, yes, but based on the same principles, all *organisms* are related to each other.

If each monk or nun, or any person for that matter, follows his or her family tree back far enough, not only will it be found that all humans are related, but if one keeps going back in time, we find that humans share a common ancestor with other primates, and then still further back, all mammals share a common ancestor, and then all animals share one, and then all plants *and* animals share one, all the way back to the mother of all organisms, some kind of single-celled creature —the original form of life.

Extending the same logic this far back is a major obstacle for the acceptance of evolution among many who hold humans as distinct from and above all other species. How could we be related to monkeys, much less to parrots or bacteria? A major societal and cultural problem, perhaps, but is it really such a conceptual leap? Proposing a common ancestor of all living things is no more than an extension (albeit a dramatic one) of those three core principles of evolution: natural selection, adaptation, and relatedness in time.

For Buddhists, this is perhaps less of an obstacle. In *How to See Yourself as You Really Are*, the Dalai Lama says that in order to truly understand yourself, you must first understand how you are integrally connected to and dependent on all other things.[13] You are an unfolding process, the sum of cause and effect of interdependent processes deep in the past and forward far in the future.

All of the scientific research of the last 150 years since Darwin published his ideas is consistent with his basic ideas. There have been and will be arguments based on new experimental data and new interpretations of those data in the scientific community concerning the finer points of evolution, but the core principles have only been reinforced.

Merely extending the principles of evolution in the Tibetan/Han Chinese example allows the argument to move from relatedness within one species to relatedness among different species. Let's do the exercise.

Certain important subtleties of oxygen use evolve in (are selected for) human populations living at high altitude for generations. Because oxygen is absolutely vital for all aerobic organisms' survival (it allows us, our mitochondria actually, to effectively produce energy), Darwin would predict that strategies for obtaining oxygen, transporting it within the body, and incorporating it into biological processes would be generally similar, not just among all humans, but among *all* aerobic organisms. He would be right. Once nature "figured out" these processes in the original aerobic organisms (bacteria), she conserved many of their components.

Hemoglobin is the protein that carries oxygen in aerobic bacteria and in Tibetans, all humans, and all aerobes. Not surprisingly, much of the genetic code, size, and molecular structure of this molecule has been conserved over these millions of years of evolution.

Darwin would be especially happy with all this new information for at least three reasons. First, as noted, genes provide a physical substrate on which his proposed concept of natural selection can act. Second, the fact that living organisms all share the same biochemical components and put most of them to use in similar ways is one of the strongest arguments that all organisms are related and share a common ancestor. For example, all organisms need to make energy (at least occasionally) in the *absence* of oxygen. Virtually all of them use the same process—glycolysis—to do this, and with the same or very similar proteins. And third, DNA/protein sequences, like those of the molecules carrying out glycolysis, provide a wonderful concrete measure of relatedness—to complement the fossil record and other more traditional methods of classification. Evolutionary biologists build increasingly precise family trees based on these sequences.

Each cell of any one organism has virtually identical DNA, but as we see in the next chapter, slight differences can have a big effect. Changes occur in DNA either during the gene mixing of meiosis or as caused by mutations induced by environmental factors from within or outside cells. Changes in DNA code can account for changes in gene function or regulation. Changes in the DNA of germ cells—sperm or egg in sexually reproducing animals—are passed on to the next generation. Natural selection "acts on" the DNA; that is, changes in DNA (and thus gene code) that give an organism an advantage are maintained and passed on. Genes, in collaboration with the environment (in ways explored in more detail in following chapters), account for traits. Similar traits are often shaped by similar genes. Changes in genes can change traits.

Now this molecular information is used to demonstrate relatedness. The degree of DNA and protein sequence similarity facilitates the understanding of (a) *how related* individuals or species are, (b) *in what ways* they are different, and (c) roughly, *when* the changes occurred in evolutionary time.

So far, then, here's the story: evolution, a conversation between environment and genes, continually fine-tunes and rearranges to create the diversity—starting from similarity—of life on earth. Such changes occur via the genes and the traits they encode that allow organisms to thrive. We can follow the natural history of these changes using DNA as a "marker" for the degree of relatedness between organisms and species.

OUR GOAL in Dharamsala is to learn each other's ideas and cultures, to work our way toward or perhaps simply stumble upon new insights, new ways to decrease suffering in the world—*not* necessarily to search for how modern science fits with Tibetan Buddhism; however, some ideas and approaches *do* integrate more easily than others. Reincarnation is surely one of the most powerfully different and distinctively "nonscientific" concepts imaginable. Yet reincarnation is at the core of the monks and nuns' Buddhism, so much a part of the fabric of their lives and thinking that it is difficult for them to fathom anyone questioning it. Questioning reincarnation for a Tibetan Buddhist is like questioning evolution for an American scientist. For a long time in our project, we leave it at that.

A vital part of our project was the weekly coffees Konchok and I (and sometimes his monastic brethren) would have every Friday for the years they were at Emory taking classes. We would reflect on science, life, and how what we were doing was changing us—our own small-e evolution. Ideas coalesced and things happened on these Fridays. It was during one of them that this book was born. Another time, an older couple approached—a couple I had noticed in the café watching us on other mornings. They politely interrupted our conversation on genetics, alleles, and cells by placing several big black-and-white photos on the table in front of us.

When she was a teenager, the woman had been on a mission trip with her family in northern India; her father was a doctor. The pictures she had were of a visit the Dalai Lama had made to her father's clinic in 1959, soon after he had left Tibet—young and smiling with his trademark spectacles. There was the woman's father and her own teenage self. And in the Starbucks a half-century later she and the monks were crying.

Over another coffee, on another day we were discussing a paper the monks were reading for their evolutionary biology class. Konchok eventually became, unmonklike, addicted to coffee like many of his fellow Emory students, some of whom actually sent it to his monastery after he left our college, until he finally had to ask them to stop. On this day Darwin's concept of the common ancestry—evolutionary brotherhood—of all organisms came up. Suddenly, we saw that this idea could jibe conceptually with Buddhism, specifically with reincarnation, of all things.

Tibetan Buddhists believe all living animals are sentient beings, and any sentient being can be reincarnated as any other. Sentient beings are brothers, related and equal within the possibilities of reincarnation.

While evolution and reincarnation are clearly concepts from different cultural universes, belief in reincarnation may actually make it easier for Tibetan Buddhist monks and nuns than for the average American to understand and accept that humans and monkeys, dogs, lizards, and fish are biological kin.

EARLY IN THE Dharamsala mornings, before teaching, I would rise with the sun and foggily climb the uneven dormitory stairs to the roof. Worn into the middle of each step are gray patches with remarkable resemblance to mitochondria, the tiny organs in my cells harnessing oxygen to crank out the energy necessary for my movement. Up there on the roof in that subtropical forest at the base of the Himalayas, it was still cool, and the diversity of birds astonishes—all of them clearly related in form and function, but yet dramatically different. Darwin's "endless forms most beautiful and most wonderful": parrotlike creatures with two-foot-long frilly tails, ubiquitous cawing crows, great bronze eagle-like things, and other birds whose brown-and-white patterns shift with the flap of their wings as I gaze at them from above.

And other flying things from up there. Later in the day, the airplane flight from Delhi that often has to turn right around and fly back without landing because of the small size of the plane and the runway, and the swirling winds off the Himalayas. Then at night, convergent evolution exemplified: a giant fruit bat swoops by in the dark as if from some bad late-night horror flick. Convergent evolution because wings were selected for—in birds and the bat, who is a mammal—entirely independently.

How do new species evolve from old ones?

As has been famously demonstrated with some other bird species, the finches that Darwin studied in the Galapagos Islands, enough small variation can accumulate, enough adaptations to different environments (say, finches originally of the same species, but now on different islands), that eventually two populations change to the point when they can no longer reproduce with each other—and thus become new and distinct species.[14]

Other times, a change in a trait or set of traits might lead to a dramatic shift and new possibilities for the organisms with the new trait(s) —another situation in which new species might evolve. Examples of such changes include those that led to organisms obtaining the ability to walk upright on two feet. Like the other genetic changes discussed, these affected traits—in this case, presumably in muscles and bones associated with the hips and pelvis—happened to give organisms an advantage in their particular environment. Dozens of different theories attempt to explain how bipedalism might have evolved, in what environments, and what advantages it provided; but however it happened, it is clear that bipedalism represented a kind of quantum leap, so to speak, in evolution, toward new species.

Similarly, mutations that allowed language, giving humans the physiological capacity to talk, represent another quantum evolutionary leap. Clues about one such mutation are found in a modern genetic syndrome in humans that causes inherited speech and grammar defects. By analyzing a gene responsible for this syndrome, researchers suggest that across evolutionary time a one or two amino-acid change in a single protein between humans and other primates resulted in a physical change in the architecture of throat anatomy, helping allow speech and the eventual evolution of language.[15] This suggests that the common ancestor of *Homo sapiens* and other primates split into different groups, one or more of which experienced this or a similar mutation that was key to providing the capacity for speech. The change, more than likely together with other changes, gave future *Homo sapiens* a huge advantage (some argue that human language allowed the evolution of human culture) to manage and survive in whatever environment they found themselves. This group eventually became a separate species. This is macroevolution.

Our point here is that *macro*evolution—evolution of new species—is really no different from *micro*evolution—evolution within species; the same core principles Darwin postulated hold true in both cases. It's just that the results of the former happen to be more striking. In the world of science, then, humans are a product of the same biological processes as, and are indeed intimately related to, all other life. From the broad biological view, as humbling as it is, humans are just another species.

EYES ARE POWERFUL, "windows to the soul," complex, knowing (figure 3.3).

The Buddha's eyes are ubiquitous in Tibetan Buddhist monasteries, temples, and sacred art. They're often, as in the figure, missing the rest

FIGURE 3.3 Buddha's eyes.

of the Buddha—alone, penetrating, compassionate eyes, looking out on the world.

In the history of thought about evolution, eyes also shine—as powerful metaphorical and literal exemplars.

Whence eyes? A final, big tension point among skeptics of evolution: How could complexity evolve? What good is half an eye? Or half a wing? Or along the same lines, how do complex biochemical pathways evolve to make, for example, DNA? What good is a pathway that only makes something that eventually will, millions of years later, evolve into DNA?

William Paley explored the eye as an example of what is now called intelligent design. He wrote in his *Natural Theology* (1802), effectively applying the biological knowledge of the time, that the complicated optics and brain connections of the eye could have come about only if preconceived by a designer.[16] In fact, back when Darwin was a college student at Cambridge, he read and was quite convinced by Paley's arguments.

In his compelling 2007 essay, Francisco Ayala argues that Darwin's famous *Origin of Species* is in large part an explanation of a design without a designer. "Natural selection was proposed by Darwin primarily to account for the adaptive organization, or design, of living beings; it is a process that preserves and promotes adaptation." It's all just natural selection—not even "progress"—just what works at that time in that place. As Ayala puts it, natural selection is "creative although not conscious."[17]

A classic challenge to evolution is how do we get an eye from no eye? Dan-Erik Nilsson skillfully walks us through the Darwinian logic, using natural selection alone as driver. Like many good evolutionary biologists, Nilsson combines observation of fossils and currently living organisms with virtual and actual models.

First, the molecular similarities of eyes across evolutionary time are consistent with Darwin's ideas. As amazing as it is given the vast differences between fruit-fly eyes and human eyes, genes similar in sequence are required for development of the eye in both humans and fruit flies. This is already strong evidence that fruit flies, humans, and their eyes evolved from a common ancestor.[18]

Remember: anything that gives an organism an advantage in its environment is favored by natural selection. Sensory organs, and their ability to gather information from the environment that results in specific behaviors, clearly provide such an advantage. Nilsson argues that eye evolution is tightly wrapped up with the evolution of visually guided behaviors. This is analogous to the physiology of the throat having to evolve before the possibility of speech can exist. The changes required to allow light sensing must have happened prior to any resulting behaviors that evolved from that ability to sense light.[19]

Nilsson connects major vision-related behaviors to four "key steps in eye evolution": (1) the ability to detect light at all, (2) the ability to detect that light and where it is coming from, (3) low-resolution vision, and (4) high-resolution vision. These four steps, he shows, correlate well with the evolution of the sensory structures needed for the behaviors. Nilsson's models take into account variables such as the speed that sensory information is processed, light detection accuracy and angle, and light intensity.[20]

The sun is a major part of the life of many of earth's organisms. Being able to detect light at all creates the capacity for evolving different behaviors related to time of day, or in the case of aquatic or burrowing animals, to how deep an organism is in the water or earth. Also, detecting shadows allows an organism to avoid predators or seek out or avoid sunlight. A single cell with light-sensitive receptors in its membrane can carry out these behaviors. Squid cells monitor that organism's bioluminescence; however, such detection can only sense very dim light.

In the second step toward evolving an eye, if the biological light detector is shielded on one side by a particular structure or pigment, the detector can determine the *direction* of the light it is sensing. Now the organism has the capacity to move directly toward or away from a light source and to pinpoint movements of predators or other objects. Usually, this is accomplished by combining pigment-containing cells and photoreceptor cells, with the photoreceptors present in stacked cell membranes. Such arrangements are found in the larvae of many sea creatures and in crustacean eyes, and they are sensitive within the range of light found in deep water or during late dusk.

To detect higher-intensity light, Nilsson's models demonstrate that special structures are needed. In his proposed third step in eye evolution, light-detecting cells are cupped in some way. Cupping allows low-resolution vision, because the organism can now detect light differences in different directions across the cup. Light detected by the cells is received at different angles and at different times across the cup's cells, allowing potential complex behaviors such as monitoring one's own movements in relation to the environment. This can lead to refined speed and direction control and increased capacity to locate a safe environment and situate one's self in that environment. Even more, the organism can now potentially detect and communicate with other organisms—whether they be mates or offspring, predators or prey. Indeed, some flatworms have one pair of cupped eyes with many receptors in each; others have many cupped eyes with single receptors —two different solutions to the same problem.[21]

The last step, allowing high-resolution vision (while retaining low-resolution capacity), requires the specialized structures of the first steps—membrane stacking and cupped cells—plus another change, one that allows the incoming light to be focused. Such a combination has only evolved in vertebrates, cephalopods, and arthropods. Nilsson shows that if the center of a light-sensitive patch of cells is constricted, incoming light becomes more focused on the backdrop (i.e., the retina), and light detection improves. In fact, this is what occurs in the modern-day sea snail, the chambered nautilus (figure 3.4); however, such an eye significantly decreases the amount of light detectable.

Another solution that also evolved focusing optics, yet does not restrict light detection to such a degree, can occur by modeling what happens when the light-detecting cup is covered with two sheets of clear cells, otherwise known as a lens. When Nilsson models a lens with two sheets of transparent plastic, a lot of light still shines through to the primitive "retina" behind the lens. If he then adds water between the two sheets of clear cells (represented by plastic), making the lens bulge out and become rounder, the image on the retina gets sharper. The

FIGURE 3.4 The chambered nautilus evolved one way to allow high-resolution vision.

more water he adds between the sheets, the rounder the lens becomes, and the sharper the image; this results in an eyelike model very similar to human eyes.[22]

Each step of eye evolution and associated behaviors probably evolved from the previous one. In existing organisms representative of each step, an increasing amount of information is processed more quickly by these organisms' nervous systems in moving from the first to fourth step. "Visual systems evolve through sequential acquisition of new and gradually more demanding tasks for new behaviors."[23]

None of these light-detection steps in evolution happened overnight. Like most complex evolutionary changes, they probably emerged from numerous smaller improvements—in this case in light detection—each of which provided selective advantage and many of which can also be identified in living organisms. The changes built into Nilsson's steps—such as membrane stacking, a patch of cells becoming indented or cupped, or water between cell layers—are not unlikely or especially startling on their own given the environment and the millions of years of time such innovations took.

An important point Nilsson notes is that much of this evolution might have started out being selected for other functions, and only later became co-opted for vision or its refinement. For example, membrane stacking occurs in sponges and other organisms that have no apparent light-detection ability. So membrane stacking probably evolved prior to light detection, was being used for other purposes already, and then only later became "useful" for light detection. In another example, some existing cupped light-sensory organs are covered with transparent material that appears to be only for protection of the organ, as they do not focus light. Perhaps some of these protective materials later evolved to focus light, that is, some of them just happened to be able to focus light, gave that organism an advantage, and thus were selected.

This is just one example of the evolution of a complex structure, and the point is not to say this is specifically how the eye evolved, but to show that following and modeling the logic of Darwin's principles *can* result in stunning complexity.

THESE THREE major points of tension in teaching and learning evolution—humans as related and similar to all organisms, the evolution of complexity, and the origin of life—are common to our classrooms in both the United States and India. In thinking through how to best translate evolution for the monastics, it becomes clear that we can also apply what we learn to our class back in the United States.

Indeed, the more we work with the monks and nuns, the more we see such similarities and resonances. In Dharamsala, the assumptions are, of course, different: with monks and nuns, we must begin our classes at different places than we do with our students in the United States, looking for different connections and entrees into topics.

Back in the United States, an underlying assumption among scientists and educators is that when students learn enough information, they eventually will "see the light" and accept evolution and other scientific concepts, while rejecting religious or other nonscientific ideas. Based on this assumption and the thinking that their job is to "just teach the science," many of my colleagues in science and science education rarely address science as part of a religious, much less an ethical or even societal context (beyond perhaps the requisite nod to human health).

In past years many people in the West had the idea that if you were a scientist, you should not be religious, and if you were religious, you should not be a scientist. Some people think that religion and science cannot or should not be in dialogue or benefit from mutual exposure to one another, because both cultures have come from different sources and have developed independently of each other.

This sort of ideology and viewing the world through a single-focus lens is common in many of the people within Tibetan communities, mostly in aged people and especially in senior monastics. Many of the senior monastics and aged people say, Why do Buddhist monks study Western science? What is the purpose of learning Western science? Mutual exposure might devastate our culture, tradition, and religion.

The main reason there is conflict between science and religion is that they lack profound knowledge of each other. There are a number of misunderstandings between these different societies and communities. It might be because people have not found appropriate methods to engage in discussion on different views between traditions and cultures.

Attitudes cannot be quickly transformed or replaced. Without examining critical views or breaking cultural barriers, it is very hard to observe the truth of the reality of others. It is also difficult to realize the benefit and deficiency of your own tradition. The best way to examine critical views of others and to gain from the understanding of others is collaboration. In order to collaborate, it is important to provide opportunities to mutually explore, to learn and share from each other.

How about learning and attitude shifts in the other direction, among Western scientists and science educators?

When we face a roomful of Tibetan Buddhist monks and nuns whose life is religion and who have invited us into this life and challenged us to demonstrate to them what science is and why it should matter to them, we can hardly avoid addressing and integrating these issues. But should we approach our American students and science teaching any differently?

As we have seen, teaching both the monastics and our students back in America gives rise to surprises—unknown biases, stunning congruencies. The pedagogical trick is being open to those surprises, hearing them, and using them to teach—not to try to convert either group, but instead, to facilitate their appreciation of how scientists think and approach a problem, what conceptual frameworks scientists use, whether or not the students' preconceptions are consistent with these, and how science can enrich and challenge their thinking.

My students at Emory are among the best in the nation, destined for leadership in the next generation of physicians and scientists. Who exactly are these students? When we look carefully at them, we see a mosaic of the world's cultures and religious beliefs—a mosaic also reflected throughout the science research laboratories of America's universities. In a typical twenty-first-century course in a private American university, the students include, in addition to those whose families have been in the United States for a century or more, first- and second-generation Koreans, Russians, Chinese, Indians, Ethiopians, Pakistanis, Somalis, and other Africans, Asians, and Middle Easterners. Not surprisingly, then, these students represent nearly all of the world's major religions: many variations of Islam, Judaism, Hinduism, Buddhism, and Christianity, as well as Jainism, Zoroastrianism, and others. While their personal views and those of science are perhaps not as divergent as those of the Tibetan monastics, their beliefs and preconceptions are certainly more diverse. These students also enter our classrooms with many ingrained ideas that seem inconsistent with science.

When I informally ask a typical upper-level biology class how many accept evolution, nearly all hands go up. But when I ask how many think something besides evolution alone accounts for modern humans in their present form, more than half the hands are raised. Science educators ignore such beliefs at our peril. As with the monks and nuns, these beliefs provide us, at minimum, with a pedagogically powerful way to engage our students, to get them to think and learn better. More important, these students as physicians and scientists in their daily professional lives will and should deal with issues that engage religion, ethics, and belief. Why pretend, as is the common practice, that these issues do not exist and are not relevant?

A recent molecular biology textbook I reviewed is typical; it includes a striking account of how researchers created from scratch a new virus-like particle that never before existed. The particle was then injected into host organisms to study what would happen to it and its host. The entire discussion had not even a passing reference to the potentially disturbing ethical implications of such work.

Teaching evolution is perhaps the clearest example of the need to engage biology in a greater context. Evolution, a conceptual framework that explains from whence organisms come, how and why they change, and how and why they are related, and a framework consistent from the molecular to the population level, is powerful, important, and essential to understand—certainly for scientists and physicians, and probably for everyone.

Evolution can get people, even monks and nuns, riled up. And that is the point—another reason it should be taught and taught well. Discussion of the teaching of evolution is rich in subtleties, which does not make for good sound bites. Humans—ironically, perhaps, because of reasons related to evolutionary adaptations for survival—are not always good at subtleties. When we hear a loud noise in the bushes in the middle of the night, we run; we do not often stop and consider that it might be a *blind* tiger, or just a friendly owl, or the wind. It is safer and easier to run. But as with Kalsang's tea experiment, the subtleties are important.

The argument for identifying and using tension points to teach evolution does not mean scientists should teach creationism or any other religious belief system. The first and most important step is not to ignore the elephant in the room, but rather, as Dobzhansky does, to name and engage it. This simple naming may not sound like much, but it makes a difference, and not just with monks and nuns.

Rachel, a former student, came to Emory to major in biology and become a physician. She was raised in Tennessee in a religious, conservative environment. Rachel's mother was a biology teacher, and Rachel came to Emory with a strong science background and a strong religious faith. When she decided to drop science from her studies, Rachel told me it was in part because whenever she entered a college science classroom, and in a number of other university venues as well, she felt her faith was either ignored, dismissed, or ridiculed. When students hear points of tension explicitly (and respectfully) mentioned, especially as entrees into learning, they report that they tend to relax, listen, and open up.[24]

A hidden but striking benefit of engaging students at this level is that they *learn the science better*. In a controlled experiment at a Wisconsin high school, students learned the science of evolution better—as compared to the control class that was taught the material in a traditional, didactic fashion—when they read and discussed original texts of Darwin and others positing non-Darwinian views of life's origins and history. The students compared and contrasted the different views and tested them with the logic of experimental science. The whole exercise was done without elaborating on or condemning personal views, but the students learned evolution and on their own determined which ideas could be scientifically investigated and how.[25] This is exactly what we are attempting with the monastics. Dobzhansky would approve.

CHAPTER FOUR

Altitude and Attitude

*If you want to understand the causes that existed in the past,
look at the results as they are manifested in the present. And if
you want to understand what results will be manifested in
the future, look at the causes that exist in the present.*

BUDDHIST SUTRA

hat makes us who we are?

Stories.

Tenzin's story: He says it was the many pairs of socks and the plastic bags from his parents that saved him. It was just about all he carried, other than the clothes on his back and the small roll of money his father tucked into his shirt pocket at the last minute.

One of our project translators, Tenzin was fifteen when his family said good-bye to him late on a cold night. They did not expect to see him again. They handed him over to smugglers, who packed him into the back of a truck in Tibet with many others he didn't know, buried him under wood and bricks, and told him not to make a sound.

The next days and weeks of hiding out and escaping over the Himalayas from Tibet to the safety of Nepal are a blur of night walking, scaling some of the highest peaks in the world, and sleeping in caves during the day. He recalls finding the bodies of fellow countrymen from failed past journeys frozen whole in the snow, and seeing the frostbitten toes and fingers of cotravelers. Every day he changed his socks and covered his feet in plastic before sleeping.

The Tibetans in India are a people in exile, a people disconnected. Many in our Dharamsala classroom have stories similar to Tenzin's. These stories are an elemental part of what makes them who they are.

Stories like these are carried from generation to generation and weave the fabric of our memories.

What makes us who we are is a driving question of Buddhism and of modern science. It seems we all seek answers to this question.

Biology is a deep part of our stories. Nothing happens in us without our cells and genes having a say, responding, translating. Whether getting hungry, eating, digesting, or walking, loving, thinking, or having a religious epiphany, everything passes through and changes our physiology, up and down the steps of our own personal living staircase.

This does not exclude God, nor does it reduce humans or other organisms to mere piles of molecules. It's simply a fact that for us to sense and respond, to act, to create or refuse to create, our physical parts must be engaged. Together in a profound, sometimes subtle, and continuous conversation with the environment, our genes both tell and help create our stories.

Many stories, like parts of Tenzin's, are tragic, broken, or buried. The Dalai Lama (a man whom Tibetan Buddhists recognize as the fourteenth reincarnation of a long and epic story) imagines that the integration of his story with that of our genes and cells might help ease, heal, lighten, and enlighten our stories, as we hear from him in chapter 8.

DURING OUR TEACHING in Dharamsala, we often take a break with other teachers and some of the monks to drink chai and relax for a moment. Chai is the drink of choice among the Tibetans without the yak of their homeland around to provide them butter and the tea made from it. Each day we have class for ninety minutes, then break for thirty minutes for chai, then ninety more minutes of class before lunch, followed by another ninety minutes of class, more chai, and then a final ninety minutes of teaching and learning.

Sometimes answers to the most innocent of questions hit hard. Small talk over chai: Where are you from? What do your parents do? Sangpo, one of the monks, has not seen his mother back in Tibet in twenty-five years. So he and his brother in India arranged a Skype conversation with her. The call took weeks to arrange, back-and-forth negotiations via mail, cell phone, and Internet cafés. When the moment finally came and the connection was made, Sangpo and his brother sat in a café in India and their mother in a café in a Tibetan town to which she had traveled from her village. The first thing Sangpo's mother asked his brother: "Who is that monk sitting next to you?"

A solid grasp of genes, the environment, and the interaction between them is valuable in order to fully appreciate these stories; what they can reveal about past, present, and future; how they speak to the question, "What makes us who we are?"

What are genes, and how do they work? What do we mean by environment? How are genes affected by the environment and vice versa? What is the impact of their interaction and what are the related emergent properties, up and down the Living Staircase? Let's explore these questions through the lens of one particular trait notable in Tenzin, Sangpo, and many of their fellow countrymen: resilience.

Unfortunately, many Tibetan people suffer trauma at an early age. This may change their personalities, and these changes may sometimes be passed on to the next generations. Tibetan people have experienced losing their country, and many were forced to flee as refugees. Tibetans in exile started escaping to India after the Chinese occupation in 1959, and since then Tibetans have continued escaping to India for many reasons. Many of them suffered trauma because of separation from parents and relatives, and some of them have been through torture; others faced desperate conditions.

Everybody is different, and so everybody has different ways of handling situations. The tough part is actually dealing with problems by accepting, letting go, and moving on. Religion and culture, social conditions and environment, genetics and epigenetics all play big roles in our response to and resilience in dealing with trauma.

Tibetan people in exile have been suffering from trauma, but rates of psychological distress are extremely low, and that is because Buddhist philosophy has influenced Tibetan habits so much.

Buddhists think about the doctrines of karma and impermanence, which help them to cope with problems or help them to get rid of trauma. For example, generally Tibetan monks and nuns meditate and study the philosophy of scriptures, and lay Tibetan people practice Buddhism through devotional practices such as prostrations, making offerings, and circumambulating stupas and temples.

Within the Tibetan people, it is rare to see the impact of parents' distress on their kids because parents, relatives, and the community have many ways to handle distress. Parents, relatives, and community have strong Buddhist influences and lead traditional practices to create a safe environment.

Tibetan people often think positively, and they remember what it says in Buddhist texts. Every evening we recite a prayer in the monastery that says that even if someone hurts you very much, do not apply evil thoughts to that person; furthermore, one should help him or her. A negative feeling toward others is the same as hurting yourself. Negative feelings are like holding a flame to be thrown at someone else, but the person holding it gets hurt. It only makes things worse to respond to a negative feeling with another negative feeling.

Our monastic classroom is a veritable testament to resilience—the capacity to bounce back, overcome, and move forward. So many of the students or their parents have stories similar to those of Tenzin and Sangpo. At least this small sample size appears biologically—physically and mentally—strong and resilient. Perhaps *because* they are in a monastery or nunnery they are more resilient; perhaps resilient young people are more likely to go into a monastic institution to start with; perhaps both things are true.

Whence resilience? The monks and nuns in Dharamsala; all Tibetans, Nepalese, Indians, or Mongolians; all humans; and probably all organisms are more or less resilient along a spectrum. Why? What makes one more resilient than the next? Why is one Tibetan who escaped now fully and effectively integrated into society, and another, who may even have been in the very same truck Tenzin was—freezing and terrified under stacks of debris, depressed and distant ever since?

The emerging answers to these questions are fascinating, complex, and far from fully understood. They connect to the gene-environment conversation at the level of these *individuals* within their lifetimes—their age, sex, and genetics; their experiences before, during, and after their escape; how they view the experience they went through; and with whom they traveled.

In the previous chapter, we explored how the gene-environment conversation at the *population* level results in adaptation and new species across evolutionary time. And as we see in the next chapter, when resilience at the individual level is considered from the broader *ecosystem* perspective, striking evidence exists that human resilience —rather than being exceptional—is actually a normal phenomenon. Indeed, psychologist Ann Masten refers to human resilience as "ordinary magic." This resonates; at some level, *all* of life, all of evolution is by definition about resilience, about how well organisms thrive and cope. Problems result when the different components of resilience —brain development and cognition, parent-child relationships, emotional and behavioral regulation, and motivation for learning and engaging the environment—are negatively affected.[1]

Mental perceptions appear to connect to and emerge from genes and cells in ways biologists never imagined. Back in America, I often discuss in my courses a study in which scientists compared the biology and psychology of mothers under constant stress (providing care for their chronically ill children) to "control" mothers of the same age and economic status living relatively normal lives. Intriguingly, chronically stressed moms who viewed their experience as stressful had cells that were biologically *older* (as measured by determining the length of special sequences at the ends of their chromosomes) than their chronological age. That is, stress seems to literally age them faster. But moms in the same situation who viewed their lives and roles in their children's lives more positively had cells with matching biological and chronological age. That is, echoing our mind-body musings in chapter 1, perspective on life appears to profoundly affect biology, resulting in a striking difference in resilience.[2] Again, the question: why?

Both conversants—genes and environment—are constantly changing and responding to one another. This means who we are, our stories, are constantly changing. We are biologically different people than we were yesterday or will be tomorrow.

CERTAINLY, Tenzin's environment as he braved the Himalayas included the "big environment," that is, the arctic temperatures, wind, high altitude, sunlight, darkness, thin air, snow, and ice. As in our discussion of Tibetans' adaptation to high altitudes, it is clear the story of the past is woven into the sequence of our genes over evolutionary time. Particular versions of genes allowed more and more robust survival at high altitude and were kept in the population and passed on from Tibetan to Tibetan.

Recall that different versions of the same genes are called alleles of those genes; most all organisms of the same species have the same genes, but different alleles of those genes. The long-term, gene-big environment conversation establishes a foundation across evolutionary time—the foundation on which the conversation continues on a short-term, individual basis.

Thus, the particular alleles of Tenzin's genes that happened to exist in his pre-Tibetan ancestors give him physical resilience at the extreme altitudes of the Tibetan plateau. These alleles help Tenzin use the scarce amounts of oxygen available in his environment without his getting altitude sickness. So, such genes are "under strong selection" in the Tibetan population because if these genes change, it dramatically affects fitness (physical resilience). The environment, in this case specifically the altitude and concurrent low level of oxygen, powerfully "select for" traits that allow one to efficiently use the little oxygen that's present; the selection is powerful, because if organisms cannot effectively use oxygen, they will die out.

Even here, it's not a straightforward one-gene/one-environment story. Recall that researchers found dozens of genetic differences between Tibetans and lowland Han Chinese. Undoubtedly, changes in different genes' sequences have similar or additive effects on fitness at high altitude. Not all Tibetans live at the most extreme altitudes. In fact, those who have migrated to the lowlands of south India, like most of the monks in our project and many other Tibetan exiles, will no longer have their "altitude resilience" genes under selective pressure. Thus, such alleles may well be lost in future generations.

Is it a coincidence that a people with such physical resilience are also so mentally resilient? Konchok talks of the influence of Buddhism and Tibetan culture. Regardless, without his genetic foundation of *altitude*-resilience genes, Tenzin would have been unlikely to reach the point of developing *attitude* resilience. As he fled, Tenzin was in a changing environment of temperature, light, and altitude that shaped his story. But also, as he ran, hid, and lived, he was at different times and to varying degrees afraid, worried, asleep, dreaming, and hungry. And he and his genes and cells responded to and translated the stimuli of these emotional or physical environments to varying degrees in different ways, ways that in turn altered his environments: put on more socks, eat, think "good thoughts," focus, or meditate. His adaptive capacity to do these things depended and depends on the genes he received from his parents and the experiences he and his genes had previously.

Some people think about the concept of impermanence to cope with problems. They consider that problems come and go, and are not eternal. To practice impermanence, people think of how plants look beautiful, fresh, and alive in summer. However, gradually the seasons change and leaves fall on the ground.

As a child, as the best way to cope with distress, I played with friends and did not ruminate on a particular event.

I went through separation anxiety when I left my home and family when I was seven years old. It stayed for many months.

I also experienced very tough living conditions for more than four years after I went to India. I clearly remember that when I was in first grade I owned only one pair of pants for more than a year. When they were dirty, I could not wash them in the common bathroom because I did not have another pair to change into.

Every weekend I used to go to the nearby spring, which was a little bit solitary and almost one mile from the hostel where I lived. I washed my pants and stayed until they dried. During the winter, sometimes it took almost the whole day for them to dry. I was not able to make it to lunch on time, and I spent many days without lunch.

During those days, to get rid of hunger and tiredness, I played games, chatted with friends, and shared jokes. When I dwelled on the tiredness and hunger, it hurt me more.

When I became a monk and started learning Buddhism, I realized there are many ways of being resilient. I learned how to accept problems, let them go, and move on.

Controlling thoughts, being compassionate, being patient, and having an open and spacious mind are the qualities of resilience. Tibetan people communicate a lot with each other; especially, elderly women gather and chat a lot, and elderly men gather and play games such as dice. Most of the elderly people keep themselves busy with circumambulating temples and stupas every morning and spending the evenings with their friends and relatives. These activities help them to divert their rumination and turn their attention away from certain traumatic events.

These are the qualities that Tibetan people have within themselves to help them to cope with trauma and other problems.

Overlaid and integrated into our genetic backgrounds, and in addition to the "big environments" we experience, other levels of environment to consider are those among and within cells. Although virtually all Tenzin's cells have the same genes, they exist in and thus respond to different *micro*environments. The genes in his brain cells respond differently than those in his heart; the genes in different parts of his brain respond differently than in other parts. That's some fifty to a hundred trillion different cells, each in a unique environment. This matters because each of these cells has a full complement of Tenzin's more than twenty thousand genes; and since genes are at the root of the biological conversation, that leaves a panoply of different environments—many of which are in flux—to respond to.

Another crucial component of Tenzin's environment related to his degree of future resilience is his *social* environment during his escape and since. Was he with other people? How many? Did he know them previously? Share stories with them? After his escape, did he move in with family or attend, as Konchok did, one of the Tibetan children's schools set up by the Indian government? Did he join a monastery? As Konchok elaborates, social support—of which religion and religious institutions are prominent examples—is a major player in resilience, physical and mental health. Seeking social support is a sign of positive coping with stress or adversity like that experienced by Tenzin. Such active coping—discussing the issue, telling your story, and being part of solving the problem—as opposed to mere passive coping—denial, avoiding the issues—affects physiology differently in the short term and creates the likelihood of more long-term effects, such as post-traumatic stress disorder (PTSD).

As we discovered in chapter 1, also vital is the interaction of time and space with environment. The internal environments of our bodies, cells, and genes change because of aging, development, and even time of day. A four-year-old would have a very different "developmental" environment in which to experience escaping over the Himalayas than would a twelve-, eighteen-, or eighty-year-old; a twelve-year-old girl is very different from a twelve-year-old boy. Think of the hormonal environment of the body of one experiencing puberty or pregnancy.

Already it is clear, then, that the questions of who we are and why one person is more resilient than another are profoundly complex. And we have only scratched the surface.

AS WE CONSIDER why Tenzin might be more or less resilient, it is also important to ask in what ways genes can be differently expressed. Recall that a gene is expressed when it is actively used as a template to encode its biologically active manifestation: RNA, and then, often but not always, protein.

As alluded to, one of the ways genetic diversity and flexibility are built into our genetic story is via sexual reproduction. Being sexual organisms provides two valuable things: *two* copies of each gene and potential *mixing* of all genes. Having two copies increases the chances of having different alleles of the same gene and therefore greater genetic capacity. In addition, during production of the sperm and egg that eventually fuse to make us, our DNA material is mixed and matched in new combinations. Some hypothesize that a driving reason sex evolved in the first place is to provide the kind of variability that comes with this mixing and matching, and thus an increase in our ability to fight disease.

Mutations, alterations in DNA sequence, can also change gene expression. Unless they occur in the sperm or the egg, mutations are not passed on to the next generation, but they might still account for differences in resilience genes. If a mutation occurred in a cell early in development, it would be passed on to the daughter cells of that cell for the rest of that person's life. So a mutation, say in a gene encoding an important player in modulating stress, could have a major impact on resilience. However, mutations are rare and usually corrected or removed; and even if they are not, their effect is often neutral or negative—certainly not something we could count on for quick adaptation within the lifetime of one organism.

NOT TOO LONG AGO, we thought of genes as fairly static. We inherited one copy of each gene from each of our parents. The genes were expressed, that is, "on" (producing RNA and protein) or "off" as needed, depending on the environment. At high altitude, on come the high-altitude genes. "Hunger genes" turn on when we're low on energy; we eat, food-digestion genes turn on, hunger genes turn off, and so on.

Such thinking may have fed into genetic determinism, the troubling philosophy that you are your genes, and therefore, if all your genes are known, if your whole genome is sequenced, we know exactly who you are. Such thinking allowed US President Bill Clinton to claim, when the first human genome was sequenced in 2000, that this gene sequence is "the language in which God created life."[3]

Were we ever surprised to learn how much more there is to the story. The truth is that saying that having in hand the human genome allows us to understand humans is like saying if we have a phone book with all the names and numbers of everyone in New Delhi, we now understand the people of New Delhi—what they look like, what they eat, their religions, who is married to whom, and how many people there like cricket more than tennis.

We were wrong in so many ways.

First, the human genome is, strictly speaking, largely not even ours! Second (and related): our genomes are dramatically dynamic, as dynamic as the environments with which they interact. This dynamic nature is evident not only in the timing and location of gene expression, but also in the creation of new versions of genes within one organism. Let's explore.

WHEN A COMPLETE human genome was first determined—when every A, C, G, and T (all three billion) of one person's total genetic code was laid out there for all to see—one weird thing was that *98 percent* of the genome didn't fit the so-called "central dogma" that has been taught in grade-school science classes for decades. (Isn't it intriguing when 98 percent of something doesn't fit a "central dogma"?) This particular dogma is that DNA encodes RNA, which encodes protein.

As it turns out, vast stretches of our genomes encode RNAs that *are not* made into protein, but (presumably) have other important functions. Other stretches of DNA don't appear to encode anything, but probably are vital for controlling or otherwise affecting the stretches that do.

Then there's the more than 40 percent of our genomes that are composed of *viral* DNA; this is the part that's not exactly ours—although it is part of our DNA, it also clearly originated from viruses. And many of these viruses can still act like viruses, that is, they replicate themselves and move and insert themselves in other areas of our genomes.

That first human genome sequence President Clinton talked about was figured out in the early 2000s and took years and millions of dollars to complete. Now you can get most of your genome sequenced in a day for around a thousand bucks. Along with this staggering innovation in DNA-sequencing technologies—in speed and efficiency—came similarly staggering improvements in sensitivity. And thus, the discovery of the microbiome, those bacterial trillions we have already begun to learn about that are our partners in life, as inseparable a part of our language as vowels are from consonants.

We dipped our toes among these bugs in our discussion of sentience in chapter 1 and will dive into them again in chapter 7 on meditation. For now, we simply appreciate the sheer capacity, the oceans of new information, these trillions of symbionts provide for us (and what we provide for them). These microbes write into our stories another hundred times more genetic information than that already provided by our human cell genomes.

What biological facts could speak more to the Buddhist principles of interconnectedness and interdependence than that we are composed of more *bacteria* cells than human cells—and that nearly half of the underlying code of our official human cells is from *viruses*? That we are as much not us as we are us? On top of that, remember that mitochondria, the engines that drive human cells, are also semi-independent organelles that probably were once bacteria themselves.

Why have viral sequences been maintained in human (and other organisms') genomes across evolutionary time? Principles of evolution say if we keep something around for generations, if it's true of many other species, then there's a good reason.

One hint is that the locations and amounts of many of these viral insertions vary from one person's genome to the next and even from cell to cell within one individual. Now this gets intriguing. What if these insertions can jump around and increase variation within a generation and from one generation to the next? Imagine they jump into the middle of a gene or into the control region of a gene and alter the type or amount of an RNA or protein that's made.

To emphasize the potential here: say that, at certain times of development under particular conditions, maybe these viral sequences are activated, or at least not prevented from copying themselves and hopping around to insert themselves in new locations, thus potentially altering gene expression, and also altering the actual genome, increasing diversity, and shuffling genetic stuff, *during an organism's lifetime.*

Some astonishing recent work on schizophrenics' brains gives you a sense of the implications here. Schizophrenia is a severe brain disorder leading to abnormal interpretations of reality and potentially resulting in hallucinations, delusions, and extremely disordered thinking and behavior. The disorder is prevalent within certain families. But there has been little luck so far in identifying specific genes that are clearly responsible.

Researchers in Japan looked at the genomes of cells in schizophrenics' brains postmortem as compared to controls.[4] They examined the prefrontal cortex, the part of the brain predicted to be affected in this disease, and they analyzed the prevalence and location of a particular viral element called LINE-1, which is known to make up more than 15 percent of our genomes and to have the capacity to jump around our genomes at certain times of development in particular cells.

Strikingly, more LINE-1 is found in schizophrenic prefrontal cortex neurons than in other cells (for example, liver cells) from the same people. Not only are there more of these hopping viral sequences in these neurons (LINE-1 elements actually make copies of themselves, which can then insert in other locations in the genome), but they tend to insert themselves near genes thought to be involved in schizophrenia. These are genes whose protein products work in neuron connectivity and signaling, for example.

The suggestion is that in developing individuals who are future potential schizophrenics, LINE-1 elements copy themselves and sometimes reinsert in the genome of cells that will become neurons of different types, possibly altering particular gene expression. This would probably happen early in development, in the embryo say, when access to DNA is, by definition, more "loose" and less restricted, because cell genomes have not yet become determined to be any one particular type of cell.

Does this happen in all developing nervous systems all the time? Based on previous studies of schizophrenics, these researchers tested in mice and monkeys (who also have similar viral sequences in their genomes) the hypothesis that viral DNA mobilization would be especially prevalent under very stressful conditions and at a particular time in early development. After mock infection of pregnant mothers, the number of LINE-1 elements indeed increased in the regions of their offspring's brains analogous to the regions of human brains thought to be involved in schizophrenia. These offspring also have schizophrenic-like behaviors.

If this story holds up, and we are only in the very early stages of understanding here, one prediction is that certain neuron cell lineages —that is, all the cells that come from one mother cell in which viruses originally jumped and reinserted themselves—will have these alterations, and other neuronal lineages will not. This was shown to be the case; in technically stunning work, another research group looked at LINE-1 in the genomes of *individual cells* in a postmortem, nondiseased human brain. The diverse patterns of LINE-1 insertions and amounts in cells in the same part of the brain fit the prediction that different future neurons must have had mutations caused by jumping viral DNA early in development so that all the ancestral cells of the originally mutated cell also have this mutation.[5]

What does all this mean for resilience? The reasoning goes something like this: if you are in an especially stressful environment, the more capacity you have for change—the more diversity potential—the better, and the more likely you or your offspring will be resilient to that environment, and thus survive and reproduce.

Viral jumping could build and increase diversity, and where better to do it than in the source of innovation and behavioral capacity—the brain? So this phenomenon might happen in such situations all the time, and the results are, according to this model, generally positive. But sometimes, if certain insertions happen in particular genes early in development and then are compounded by additional stressors later in life, as is thought to be the case with schizophrenia, bad things can happen also.

NEAR KONCHOK'S MONASTERY is a hostel for boys; they come from all over the Himalayan range—Nepal, Tibet, Bhutan, and India. A set of twins in their young teens, Ramu and Shamu, moved to the hostel a couple of years ago. No one could tell the twins apart just by looking at them, only by talking with them or seeing their handwriting.

During the classes Konchok taught for his fellow monks on genetics, several asked him about Ramu and Shamu. How would biologists explain their physical similarity, together with their differences?

Think about our discussion through the lens of the monks' question. We call them identical twins because they look the same and they have the same genomes. But, of course, we all know identical twins do not look identical for long, and now we have learned that they do not even have identical genomes. The mobile viruses in their genomes have jumped to varying degrees and into different locations, depending on the different environments they experienced; their microbiomes must also be different for the same reason, and therefore, the genomic information and what those microbiomes are doing is also different. And now, more.

Imagine that Tenzin had an identical twin who had stayed behind in Tibet. If Tenzin and his twin were reunited twenty years later, anyone who has known identical twins would probably agree that Tenzin and his twin would have become different. Few, though, would realize that our understanding of *why* the twins would be different has a lot to do with a handful of Swedes in a small obscure subarctic parish and a few Canadian lab rats minding their own sweet business in Montreal.

Överkalix, in addition to being a beautiful tourist haven tucked away on the Kalix River in northern Sweden, is a geneticist's dream. Studying the gene-environment conversation in humans is extremely difficult. A good experiment keeps everything constant except for the specific thing being measured. Very little about daily life is constant between any two people, much less among a population. Humans rarely stick around in the same city anymore, much less in the same environment where their parents or grandparents lived. And if they do, they rarely keep good records of that environment. Överkalix is different. The town is very isolated, so historically it had very little immigration or emigration. The citizens have maintained wonderful records of who married whom, who died of what when, and crucial to this story, what the availability of food was every year. Överkalix had to be self-sufficient, and thus the health and welfare of its community depended on how well their crops did.

Lars Bygren found a surprising connection between food availability

—a potential stressor—and the cause of death among Överkalixites in following generations. The correlation only held when food shortages occurred during a certain window of an individual's development: the time of slow growth prior to puberty.[6]

If your father *lacked* food during that period, you were much less likely to die of cardiovascular disease than controls. As striking, as an Överkalix citizen, if your paternal grandfather had access to an *excess* of food directly before puberty, your chances of dying from diabetes were significantly increased. Granddaughters' mortality was also associated with their grandmothers' food supply in similar ways. It seems that, as with Brian Dias's (much later) experiments with mice and smell, *an experience or sensitivity to an experience* was somehow transgenerationally inherited in a time frame much too short to be accounted for by a mutation in DNA sequence. Part of people's stories, key plotlines, were inherited over a short time.

These findings resonate with the centuries-old ideas, long ago laughed aside, of Jean-Baptiste Lamarck we discussed in chapter 3.

NOW, ABOUT THOSE lab rats in Montreal, where the story gets even more intriguing. Following from the work on Swedes, Michael Meaney and colleagues connected the dots.[7] They used genetically identical rats to directly link an early life experience, *the inheritance of that experience*, to a change in the expression of a resilience-related gene.

Epigenetics, most broadly defined, is the study of changes in organisms because of alterations in gene expression—as opposed to alterations in gene sequence. Some more refined definitions claim that these alterations can be heritable—such as the sensitivity to smell linked to a shock passed on in Brian Dias's mice in chapter 2. Epigenetic changes happen in response to the environment; epigenetics is the hyphen in the nature-nurture debate.

Perhaps the most straightforward way to change the gene expression between Tenzin and his hypothetical identical twin is to somehow switch off resilience genes in one twin but not in the other. Such an alternate expression could be in different cells at different times in different parts of the brain.

One can have the very same gene, but if it's not expressed, it's really no different—at that moment in time—than not having the gene at all.

So what matters is the *control* of these genes, the regulation of on and off. Control access to these genes and control these genes. In each nucleus of each of our cells is an entire genome's worth of DNA and the genes it encodes—feet long in each cell!—tightly bundled in very organized ways.

The DNA is bundled around proteins, called histones, that control access to the DNA. Other proteins change the histones or the DNA itself so it is more or less accessible. Once accessible, the machinery that reads the DNA and converts its code into RNA can move in. This machinery is, of course, as important as access control. Currently a bit of a controversy is raging over which is *more* important—access controllers or the protein machinery that moves in when genes are accessible. New, big ideas such as epigenetics often lead to controversy; but working through such controversies is what science is about, and does not appreciably change the spirit of our particular conversation.

In fact, we have known about one level of epigenetics—inheritance between the same types of *cells*—for quite a while. Those cells we talked about in chapter 2, the cells going through development, differentiate and become gradually more determined, or set, to be one type of cell or another. In molecular terms, this means a skin cell, for example, turns off (shuts down access to) all the genes not involved in being a skin cell and turns on (allows access to the necessary activating factors) all the skin-cell specific genes. Then when a fully differentiated skin cell divides, it "remembers" it is a skin cell, that is, that it has this specific skin-cell epigenetic state, and passes on this state of gene-access to its daughter cells, which are each "born" as fully differentiated skin cells themselves.

But here we are talking about a different level of epigenetics, the memory of gene access and thus gene expression and resulting phenotypes, passed between *generations*—this is the new, more controversial bit.

Meaney found that mouse pups who received a lot of good attention from their mothers *during a key time in those pups' young lives* grew up to be calm and resilient as compared to pups that were raised by neglectful moms. (Maternal care was measured by how long the mothers spent licking and effectively nursing pups.) This was true regardless of whether the mother was the genetic or adopted mother of the pups,

FIGURE 4.1 When mouse mothers (on the left) take
good care of their pups, a particular gene (indicated by Nr3c1)
is turned on; this gene's expression is blocked in pups of mothers
(on the right) who pay less attention to their pups, impacting
the pups' response to stress once they are adults.

so the effect had to be *epigenetic*, not genetic. Not only that, but the scientists connected this greater resilience directly to the change in accessibility for control of a specific gene.

Glucocorticoid receptor gene expression in the pups' brains varied depending on the quality of attention the pups received. This gene is known to be crucial to resilience, to stress response. Access to the gene was chemically blocked in the adult mice that had been relatively neglected as pups (on the right in figure 4.1), but not in adults that as pups had been well taken care of (on the left in figure 4.1). In addition, the *experience* of early maternal care was inherited. That is, females raised by caring moms became caring moms themselves, and females raised by less-caring moms took less care of their own pups.[8]

All of us in this world have our own particular capabilities and differences in our abilities, and attitudes are not the result of chance. Inequalities among people, such as the physical and mental ones that exist, are to some extent attributable to the environment, which is shaped by cause and effect, or karma.

Everything in the universe exists within the framework of cause and effect. Every action and every thing are interdependent, and no action or thing is produced without a cause.

All phenomena arise because of multiple causes and conditions. It is not possible to produce something without substantial cause and cooperative conditions. A seed can be considered to cause a plant to grow, and soil and all the necessities for the seed to grow are considered the cooperative conditions.

Gradually things emerge or evolve into more complex systems that have more stability and strength. For example, the brain is made of cells only, but if you just have neuron cells, then it will not function appropriately; but if the glial cells and all the other necessities are there, it will perform in unison. Therefore, I agree with the statement, "The whole is greater than the composition of its parts."

I think that the mechanism of karma in Buddhism is comparable to the mechanism of epigenetics in biology. Buddhists believe that the past influences the present, and the past and the present influence the future.

This is very similar to epigenetics. If you want to understand the causes that exist in your genes, look at the results in the phenotypes that emerge. And if you want to understand what results will be manifested in the future, look at the causes that exist in the present genes.

My knowledge of karma gives me a little bit of insight into epigenetics. Epigenetics can involve a kind of inheritance that is not actually encoded in the DNA sequence itself, but governs when and where genes will be active or inactive in the different tissues.

The power of karma you accumulated in the past determines or governs how and where you will be reborn. If you have accumulated karma to be born in a blacksmith's family, then you will be reborn in a blacksmith's family, not in a king's family. This is the same idea that you reap what you sow. It is not like you sow rice and reap barley.

There is a common example in Buddhist texts; it says if you are generous in this life, the consequences will be manifest in the next life with complete enjoyment and fullest wealth. Buddhists believe that every action, no matter how insignificant, produces a corresponding result, which is a similar notion to Newton's third law of motion: for every action, there is an equal and opposite reaction.

In Buddhism, inequalities are not only attributable to heredity and environment, but also to karma. They are the result of past actions and present doings.

Epigenetic information passes on when cells divide and maybe for some generations, but it is not permanent. Karma is also like this.

A difference is that biologists talk about how this present life affects *off-spring*, whereas Buddhists talk about how the present life affects *the next life*. A similarity is that biologists and Buddhists both think that actions in one's life affect one's later life and perhaps the future lives of others.

Kerry Ressler is an MD and PhD, a psychiatrist, and a researcher, and for a guy who studies stress, a remarkably upbeat and talented Renaissance man. We mentioned Kerry as the lead collaborator on Brian Dias's experiments on smell and shock in mice. I first met Kerry when I invited him to be a member of a symposium panel on dread. We also invited humanists and social scientists whose work touched on this emotion, as well as one of my old students, Kim, who's now a professional oboist, to perform a prelude to the program and again at intermission. No music captures dread more than the haunting, double-reed wail of an oboe.

Kerry is doing some of the most rigorous and compelling work on understanding stress. He and collaborators from around the world integrate the diverse aspects of stress we have been discussing up and down the Living Staircase. They have been following a cohort of African-Americans at Atlanta's large inner-city hospital. These people have agreed to have their genomes, cells, brains, and personalities studied in hopes of better understanding stress, specifically post-traumatic stress disorder, and thus intervening to prevent it in the future.

Kerry and his collaborators study diverse aspects of the so-called HPA axis that mediates psychological and physical resilience, or stress resistance. The HPA axis gets its name from the first letters in three of its major interacting components: the interacting Hypothalamus, Pituitary, and Adrenal glands (shown in figure 4.2). Other parts of the brain integrated with the stress response include the amygdala, hippocampus, prefrontal cortex, and reward circuits. Thousands of studies in humans and animal models combine to show how all these structures, and the hormones, neurotransmitters, and receptors they use for communication, contribute to resilience. In chapter 7, it will also become evident how the microbes that coexist within us also interact with the immune system and other components of the nervous system to monitor and respond to stress.

The genes we receive from our parents that encode the brain structures and the molecules which integrate their functions potentially

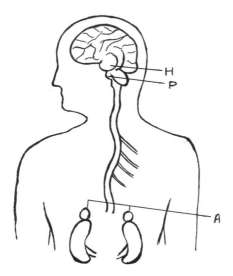

FIGURE 4.2 The HPA axis—short for Hypothalamus, Pituitary, and Adrenal glands—that helps mediate the stress response in humans.

affect our degree of resilience. These genes and their mechanism of inheritance are very similar to the "altitude-resistance" genes we discussed previously. They evolve over generational time and are conserved within the population if they offer survival advantage.

Ressler and his colleagues have found that a *single* difference in the DNA sequence controlling a specific HPA-axis gene makes a big difference in this story. Officially, this gene and the protein it encodes are called, for reasons we won't go into, FKBP5; for simplicity, we'll just call it F factor.[9]

First is the basic plotline of the stress response at the molecular level. Just as in mice, a central player is the glucocorticoid receptor protein. In nonstressful situations, this protein lives in the cytoplasm of cells, outside the nucleus (where the DNA resides). Once stress occurs, the HPA axis releases cortisol, the major stress-signaling hormone, into the blood.

Cortisol sends the "stress message" to cells throughout your body: we've got a problem; quickly make a bunch of proteins to help us deal with this! To do this, cortisol enters cells and binds the glucocorticoid receptor protein. Then this protein plus cortisol goes into the nucleus and turns on those genes needed to make that "bunch of proteins."

Now we get to the F factor. In general, it's important to eventually turn off anything that's turned on. Biology builds the off switch into the on switch. This is called feedback. Here, one of the genes stress turns on is the F-factor gene. The F-factor gene encodes the F-factor protein, which feeds back to prevent the glucocorticoid receptor from binding cortisol. And the stress response is turned off.

You can imagine how altering the basic plotlines—interfering with any of the basic personalities of our major characters: cortisol itself, its receptor, the F factor—might have something to say about Tenzin's stress response.

Ressler found that a single DNA base-pair change can shift the plotline; the change is in a regulatory region of the F-factor gene that receptor plus cortisol requires in order to turn on that gene. In some people he is studying, Kerry finds one allele that has an A-T base pair in that location, and in others a C-G base pair; he finds that the former protects adults against suffering from post-traumatic stress disorder, while the latter makes it more likely people will suffer from PTSD as adults. These connections only hold true if one suffers the trauma (as self-reported in these studies) early in life.

What's going on? Again, we see that the *timing* of the stressor is important—in two ways. First, the trauma has to happen early, and second, when it does happen, it can have an effect later. So far, Kerry and colleagues have been able to look only at that "later" within one individual's lifetime, but soon they will be able to see if the effect goes across generations, as it did with the Överkalixites.

At the molecular level, it appears that those with the susceptible allele (C-G) who suffer early trauma experience a chemical modification, an *epigenetic* change, in the part of the F-factor gene that controls its regulation. The result is less production of F-factor protein, so the receptor for reporting stress is more active, and the stress response in general is more pronounced; feedback is altered. Perhaps the baseline stress level is higher, and PTSD is more likely.

The "more likely" part is crucial to the resilience story and to the understanding of any complex trait. No single event or single gene guarantees PTSD. As we have seen, resilience is much more complicated than that. First of all, you have two copies of each and every gene, including the one for F factor. Then, so many, many genes are involved in resilience. Then there are the jumping viruses, the microbiome, the environment, how you cope, and on and on.

AS KONCHOK points out, the Tibetan Buddhist culture has many built-in coping mechanisms that other cultures may not. In fact, Kerry recently published the results of a study with Carol Worthman, another of the leaders in our project, showing that, in a group of Konchok's country-men, F-factor alleles that correlate with PTSD and depression in other groups correlate only to depression among this Nepalese cohort.[10] The reasons for this are unclear. The F factor still appears relevant here; and given the complexity of resilience, the syndromes being studied, and human populations in general, it's not surprising, and in fact is common, to find such differences in correlations between genes and disease.

Two other striking findings from Kerry's research: (1) the connected genetic changes and PTSD symptoms also correlate to differences in the architecture of brain structures in the HPA axis, and (2) the changes of F factor and its effects are true in brain cells as well as blood and im-mune-system cells, further setting the stage for our discussion of mind-body interactions in chapter 6.[11]

Why would an allele that makes you more susceptible to stress be maintained in the population? Perhaps in some stressful environments it pays to be more stress aware, more sensitive to dangers.

WE INTUITED that Tenzin in India and his hypothetical identical twin who stayed in Tibet would be different twenty years later, because their environments—in all the different versions of that word—are dramat-ically different. Now we have begun to see the reflection of these dif-ferences and their impact at the gene level. In fact, there is no more dramatic way to see, literally, this impact than to look at identical twins' chromosomes over time. In figure 4.3, the left column of two sample chromosomes is from twin A and on the right from twin B, and time is increasing in years from top to bottom. The cross-hatching represents differences in access to various genes; with time, as the twins' environ-ments inevitably become more different more often (that is, the older the twins get), so does the access to their genes, and thus the expres-sion of those genes' changes.[12] And this would be true to a different extent on different genes in different cells. Most likely, it will get easier and easier as time goes on to differentiate Ramu and Shamu.

We do not yet know which gene access marks are passed on to the

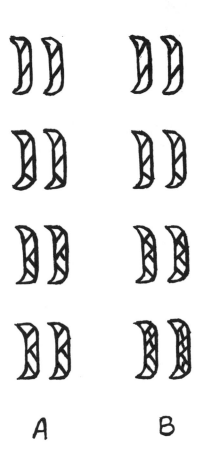

FIGURE 4.3 Sets of chromosomes from identical twins, A and B; while the DNA sequence of the twins remains identical, with time (moving from top to bottom) the different environments and experiences of the twins alter the epigenetic marks (indicated with cross-hatching) on and around these DNA sequences, resulting in differential gene expression.

A   B

next generation or why or for how many generations, although two to three generations seems to be the average. We do know resilience and its accompanying epigenetic marks are affected by the intensity, predictability, and duration of stressors, as well as the age at which they occur.

And thus, the story of how fifty trillion cells with more than twenty-thousand genes each in conversation with their environments becomes that much more difficult to wrap our minds around.

But we keep telling the stories and continue to try to understand them and each other through them. Stories themselves are a kind of social support, a kind of resilience builder. Good stories are rich with meaning and drama and lessons. Good stories are told over and over again, passed on like genes. And they change us. They are part of our environment, and thus they are part of our genes. Can our new knowledge of genomes and epigenomes help the broken stories heal? Are epigenetic marks reversible?

IS THERE A WAY to integrate students' stories into their learning? By so doing, can educators enhance learning? Remember Dhondup's insight: he studies science to understand Buddhism better, to become a better Buddhist. Since people learn better when the material is relevant to their lives, surely they will learn well if the material is seen as or is made part of their own personal stories.

Even if they see this general aim as important, teachers are often at a loss as to how to do this. The challenge is even pricklier if the story involves science and religion. And yet, for Dhondup and many Americans as well, such stories are potentially the most impactful.

Once one of the monks unexpectedly moved me to tears in class. When the monks and nuns were studying physiology in Dharamsala, we assigned small groups to each study one particular organ system. Each group had to explain that system to the rest of the class and then describe what would happen within that system in a specific life situation. Kalsang stood up. "Here's what will happen to my circulatory system when Tibet is free again," he said. "My blood pressure will rise, hormones will be released, my heartbeat will increase."

Making personal stories part of teaching and learning is especially rare and challenging, even looked down on, in science education. Nevertheless, our experience with the monks drove home the benefits of stories for learning—even in science. If the goal is for those in our classes to fully learn the science and to better understand themselves, who they are, and who they are becoming in the context of that science and in society as a whole, this is well worth the effort. Make the science learning part of the experiences that make students who they are.

One relatively simple approach with little risk is to have students shape the syllabus based on their life stories. Starting out my cell biology course back in America, I ask the students to identify the three things in their lives with which they are most concerned. After I reassure them that I am serious, in response to their nervous glances at each other, we develop a list of issues on the board. They are somewhat predictable for twenty-year-olds: stress related to school or being admitted to professional and graduate schools, boyfriend and girlfriend issues, depression, music, spirituality and religion, drugs, diet, exercise, general well-being.

Then I tell the students that none of their concerns or actions, or even their thoughts, can happen without cells. So all of the issues on our list must engage cells. I then select papers from the recent scientific literature related to the issues most relevant to the students' current stories and use those papers to shape the course and teach cell biology. A typical class has papers that fall into topics such as relationships (cooperation, sex, sociability, language, music and hearing, and religion), stress (aging, diet, and meditation), stem cells (health, disease), cell patterns and the environment (regenerative medicine, symmetry), and learning (circadian rhythms, enhancing memory).

Inspired in part by our work with the monks and nuns, we also experimented in the classroom back in the United States with a deeper integration of personal stories and science. On the first day of a cell biology course one semester, I announced that in parallel with the course we would offer a seminar to focus on the religious and ethical perspectives of the science explored in the cell biology course. I expected just a handful of students would be interested, especially given that I was announcing the seminar after the semester had already begun. *Half* the students out of fifty in the course, though, stayed after class to express interest. In the end, fifteen students met every other week during the semester. Even in such a small group, the diversity was striking: at least one student identified as Jewish, Muslim, atheist, Buddhist, agnostic, Hindu, Jain, and Christian.[13]

In the seminar, students introduced themselves through a brief personal statement telling the story of their beliefs and from where those beliefs came in the context of what drew them to this particular seminar. We used these personal stories and the discussion of them that followed as a social and intellectual starting point from which to explore issues raised in the science course from the ethical and religious perspective. What are the points of complementarity and tension, and how do we work through them?

Did religions and religious behavior evolve like any other trait? And if so, what are the implications? For example, why do so many religious rituals around the globe use incense? The Three Wise Men brought it as a gift at Jesus's birth; Buddhists call it the "fragrance of pure morality."

In the cell biology course, we discussed a protein receptor—like those described previously—that when activated decreases anxiety, how this receptor is activated by a chemical in incense, and the implications for the evolution of shared ritualistic practices and their relationship to our biology.[14] In the seminar we asked, "If a phenomenon is explained scientifically, does it lose religious value or importance?"

I had the opportunity to take Dr. Arri's cell biology course at Emory with Sangpo, another monk we mentioned earlier in this chapter. In one session Sangpo, a third monk, and an American student presented a paper in front of the whole class about how incense reduces stress and brings you into a pleasant state. They mentioned that some part of the incense binds to a certain receptor in cells and reduces anxiety, kind of increases resilience, in mice.

The incense-burning tradition is rooted in many religions, especially in Buddhism and Hinduism, as I witness very often. I myself use incense in my room every day, especially when I sit to do my daily prayers and meditation. In Buddhism, incense is mostly used during religious ceremonies and on some other special occasions. When you visit most of the Buddhist temples, you will smell the fragrance of incense all the time. We think of incense religiously for "purification."

Incense burning is also connected with modern science, especially with biology and health issues. You might think incense is more connected with spiritual activities and religious works, so is it odd to talk about incense in biology? I think not; both work together. Burning incense is surely helpful for reducing stress and bringing pleasantness. Something changes in your molecules, your biology. Otherwise, how else can you feel it?

This study in cell biology class changed my experience with incense.

After an intriguing cell biology course discussion of the latest in stem cell research, in the seminar, students explored why it may have been that George W. Bush used his very first national television address as president to discuss embryonic stem cell research and policies. Why did President Obama later reverse some of those policy decisions?

To maintain the momentum of students' personal stories and the integration of the two classes, students kept a journal. To measure the effectiveness of this two-course approach, in addition to written course evaluations, several students agreed to be interviewed anonymously following the project.

First, it was abundantly clear that the desire is there: students very much want to explore the science they are learning in the context of religion and of their own personal beliefs, and many stated that the opportunities for them to do this are rare. When we put out a call for a similar seminar to physicians and other professors at our university, nearly one hundred responded positively.

The strong interest in the seminar also surprised the students taking it and spurred learning. Cheng—quick-thinking, outspoken, and on his way to medical school—expressed a view held by many: "I was surprised so many science majors claim a religion. I'm an atheist, and I had a stereotype that all scientists were like me, but this experience opened my eyes. It means the gap between science and religion isn't as clear-cut as I thought."[15]

So simply providing safe space and opportunity might be a key and relatively easy step, so long as educators are prepared to productively harness the ideas and energies that ensue. Cheng and Gwen, a thoughtful evangelical Christian, were especially emphatic in their declaration of the need to provide these opportunities for future scientists and physicians. They stressed how important it was to have these two courses —cell biology and the seminar—linked and integrated, saying they would not have taken such a seminar on its own.

Strikingly, the students spoke positively of the impact of this project on their *knowledge of biology*. The courses helped them think "more critically and holistically," and they were now able to better consider the impact of knowledge and biology on other aspects of life and culture. One student noted: "I could see Biology in my life more prominently. I made more connections, I came up w[ith] thought questions just walking to class. I had intelligent things to say in conversation that non-science majors found interesting. This class really challenged all my beliefs across the board, and I feel like I learned a lot."[16]

As another participant said: "I wouldn't be thinking the way I am now [without having taken these two courses]. I thought cell biology would be just another science class I'd have to memorize and study for, but even studying for the [cell biology] midterm I was coming back to the religious stuff."

Sharing personal stories and weaving them into the science they were learning, and their discussions of the ethical and religious aspects of that science, strengthened the social and intellectual engagement of the seminar students. Their conversations and insight often moved beyond learning biology to more metacognitive questions about *how* they were learning. They explored the very nature of the project and their preconceptions. Several students admitted they initially gave scientific knowledge superiority over religion. Also at the metacognitive level, the students wrestled with their own role in the public science and religion conversation—now and as future scientists and physicians. How should government and professional policies affect their actions, especially if they personally disagreed with those policies? Should they perform experiments when they had ethical or religious qualms related to those experiments? As physicians, would they use therapies developed from research with embryonic stem cells if they thought that research involved the murder of embryos? How, as scientists, could they best impact relevant policies and engage these issues with family, patients, or coworkers?

Again reminiscent of the monk Dhondup and his insight, students in our two-course project did not radically change their minds, but rather changed *the way they look at things*, and perhaps grew to better understand their own minds—both surely goals of liberal arts education.

Gwen, a Christian, elaborates: "I loved everything else about biology because the more I studied it, the more I learned about God's creativity and the intricacy in his design. Biology was becoming a big part of my life, and religion was already a big part of my life. When I heard about this class, I was immediately interested. I've never taken a course that talks about religion and biology, my two great loves. I thought that it would be a good opportunity to let both topics fuse, instead of keeping them independent from each other. It was time biology finally met God."

Cheng, an atheist, said: "Science is driven on the basis of fact and empirical evidence. The more you learn about science, the less plausible any type of religion that refutes scientific methodology becomes. The 'God of spaces' [sic, 'God of the Gaps'] was introduced in our seminar class, and that theory states that 'God' is merely a tool used by people to mask the void of not knowing. Religion exists as a tool, the so-called 'opiate' of the masses. It exists to console them when science cannot. That is the connection between science and religion as I see it."[17]

THIS CHAPTER BEGAN by asking: What makes us who we are? In a way, our project teaching science to the monks and *every* course is somehow addressing this question. A key concept in teaching and learning is *transference*—the capacity for learners to transfer ideas, facts, and concepts learned in one context to another situation.

The unique experience of teaching science to Tibetan monastics in Dharamsala and in the United States drives home to us the power of integrating personal stories into teaching. Our success in Dharamsala catalyzed the transference of this approach to the initially very different seeming American university science classroom. The impact on teaching and learning is striking.

Personal stories and belief discussions are clearly not a traditional part of science education in the United States, but given the impact on the improvement in learning, application, and general scholarship, perhaps they should be. What is the purpose of learning institutions if not the integration of knowledge, the creation of new ideas, and the development of better citizens with rich personal stories?

Anxieties about attempting such approaches are reasonable, but the current and traditional science teaching structure does allow room for relatively easy experimentation. Many scientists already step into this territory in courses or in public discussions, so maybe the leap isn't as great as it first appears. For example, in seminar courses, geneticists commonly discuss the ethics of genetics or stem cells, and physicists discuss the societal impact of nuclear power. In both the Tibetan monastic and American two-course projects, we spiced things up by being more intentional and rigorous about our engagement. We linked our projects directly, in different ways—that explicitly address personal stories, beliefs, and religion—to "traditional" science education.

Stories.

In collaboration and conversation with our environment and experiences, genetics and epigenetics shape our stories through time—the resilience of Tenzin and Sangpo, the beliefs and learning of Gwen and Cheng. Accessing and understanding these stories helps us learn and develop as individuals and as a society. The more the process of learning engages and changes our own stories, the more powerful that process becomes.

CHAPTER FIVE

Ecology and Karma

Peerless One
Who, seeing the all-pervasive nature
Of interdependence
Between the environment and sentient beings
Samsara and Nirvana
Moving and unmoving
Teaches the world out of compassion
Bestow thy benevolence on us.

REFLECTIONS COMMEMORATING
THE DALAI LAMA'S OPENING OF THE INTERNATIONAL
CONFERENCE ON ECOLOGICAL RESPONSIBILITY:
A DIALOGUE WITH BUDDHISM
[other verses distributed throughout this chapter]

Interdependence is the spiritual truth
that biologists have independently discovered through
the scientific discipline of ecology.

ECOLOGICAL BUDDHISM: A BUDDHIST RESPONSE
TO GLOBAL WARMING

he connection between Buddhism and ecology is profound. As the second quotation that opens this chapter suggests, Buddhism discovered ecology before ecology did. The interdependence that is ecology (a term coined in 1866) is a central concept and value of Tibetan Buddhism. Indeed, interdependence is tightly woven with two other central Buddhist concepts: emptiness and compassion.

Mothers are held up as the ultimate examples of compassion and empathy in Buddhism. They provide for and unconditionally love their children. All living beings can be thought of as our mothers. Humans rely on other humans (and other organisms)—many with whom we never directly interact—for food, clothing, transportation, and safety. We benefit from each other's "hopes, dreams, and labor." All beings need each other for survival in our one world; we all need each other's compassion.[1]

The self is on the other side of the spectrum. From both biological and Buddhist perspectives, the self is always changing. Buddhism takes this idea one more step to say that, therefore, the self is *impermanent*, an entity empty of permanent essence—not meaningless, but always shifting and evolving. Thus all emotions, experiences, and actions are also impermanent and empty in the continually changing, moving, adapting lives humans (and other sentient beings) lead. Because of this vital emptiness, attachment is unnecessary, even problematic. The realization of this impermanence of things allows people to "develop equanimity regarding all phenomena."[2]

A key point here for the relationship of Buddhism and ecology is that letting go of the fixation on the self enables one to soften the distinction between self and others and thus to more easily behave as a compassionate part of the whole.

THE TIBETAN PLATEAU is one of the most important ecosystems on earth. (Figure 5.1 shows its location within Asia.)

The plateau's glacial ice helps provide water to nearly a quarter of the world's population.

Perhaps much of the future of the human species hinges on these glaciers in the context of the region's vast grasslands, stunning lakes, millions of square miles of permafrost, extensive forests and wetlands —home to over twelve thousand species of plants, hundreds of bird species, yak, panda, snow leopard, and Konchok and his several million nomadic Tibetan brethren.

Is it more than mere cruel irony that the resilience of the Tibetan Plateau ecosystem is being dramatically tested at the same moment as that of the Tibetan Buddhist culture?

In my home region of Phuksundo on the Tibetan Plateau, we are all ecologists. Our lives, religion, and ancestors deeply connect us to our environment.

FIGURE 5.1 The Tibetan Plateau, native home to Konchok
and most of the monastics in our project.

I remember hundreds of stupas, shrines, and stone pyramids as far as I
could see in all directions, and especially at every mountain pass. They are
indicative of the religious devotion my people have to our local deities, who
connect us to the land and our ancestors. The villagers worship and honor
them throughout the year.

In summer, when the villagers move to higher altitudes to graze their
animals, everyone gathers next to the shrine of the local god. Families
bring *tsampa* (roasted barley flour), yogurt, *chang* (local beer), and dried
fruits. The religious leaders take part in preparations for the annual ritual
ceremony honoring the local deities and their ancestors. *Tsampa* is used to
make *tormas* (offering cakes) of different animal shapes and colors (red for
wrathful, yellow for peaceful) to embody each individual deity, local god,

spirit, and serpent *naga* (evil spirit). We place these offerings around the shrines, which have been beautifully decorated with *khatas* (traditional scarves) and colored wool—of white, blue, red, yellow, and green. While this is happening, the women and girls collect firewood, juniper, and other aromatic ingredients to make the smoke offering, or purifying fumigation. Some women clean the area and make butter tea; tea and *chang* are served to everyone after the preparations have been completed. Aged people consider *chang* an energy drink.

After half an hour of rest, all the villagers participate in the ritual ceremony, which lasts about two hours. The ceremony is for honoring and requesting that the deities, local gods, spirits, and ancestors provide a good harvest, good health for everyone, and favorable weather.

After the ceremony, we sing traditional songs and perform traditional dances dressed in beautiful handmade clothes; we play games and drink butter tea and lots of *chang*. Usually parents do not serve *chang* or any other kind of alcohol to their children, but children are allowed to drink a little on this special day.

On two occasions, I was naughty and drank more than two cups; unfortunately, my mother caught me when I was totally drunk. I still feel embarrassed about this incident when I think of her.

Konchok spent his childhood in Phoksundo in the Dolpo region of the plateau, home of Nepal's most famous lake. Dolpo is on the western edge of the Tibetan Plateau, the highest and largest plateau in the world, encompassing nearly a million square miles of the Tibetan Autonomous Region, as well as parts of India, Nepal, and Bhutan. The plateau is bounded by some of the highest mountains on earth and averages more than sixteen thousand feet in elevation; Konchok spent his youth living at elevations between twelve thousand and twenty-one thousand feet.

To travel by caravan from Lhasa, the traditional Tibetan capital, where the Dalai Lama and his predecessors resided until 1959, in the south-central part of the plateau to Xining in the northeast corner takes four-and-a-half months. The new Chinese train between these two destinations makes the trip in twenty-four hours (including five additional stops); from Beijing to Lhasa takes forty-three hours by train. The route involves such extreme altitudes that the train is pressurized and sealed with oxygen provided—much like an airplane—and also has ultraviolet filters to protect travelers from severe sunburn.

This train is part of a dramatic engineering effort—both scientific and cultural—by the Han Chinese, the politically dominant ethnicity in China and the biological brethren of Tibetans. The train provides relatively easy access to the Tibetan Plateau for the first time in its history. This, together with economic incentives, allows and encourages the Han to migrate to the plateau in unprecedented numbers, with an enormous array of sociobiological implications for them, the Tibetans, and the plateau.

AN ECOSYSTEM refers to the complex set of relationships of all living and nonliving components of any given environment.

Describing the characteristics of the ecosystem of the Tibetan Plateau is like reading a Guinness Book of Ecosystem World Records, a long list of superlatives with a long list of related current and potential impacts on the planet. In addition to its sheer size and altitude, with the exception of the North and South Poles, the plateau holds more ice than any other region on earth—more than 46,000 glaciers covering over 65,000 square miles. This ice is melting fast.

The plateau and its glaciers are the source of some of the world's longest and most important rivers—including the Yangtze, Yellow, Indus, Mekong, Ganges, Salween, and Brahmaputra—that provide water to over a billion people in Asia—in India, Pakistan, China, Myanmar, Laos, Thailand, Vietnam, and Cambodia.

The region of Dolpo is made up of many unexplored beautiful lush green pastures, snow-covered rocky mountains, beautiful lakes, mountain streams, and thick forests that are spectacularly beautiful. It is a region rich in natural resources and has an ecosystem with a huge variety of living organisms. My people who inhabit these areas are strongly connected with their land, ecology, and ancestors.

Historically, we have always been highly dependent on natural resources, such as forests and pastures, because most of us are nomads, dependent on farming and animal husbandry for our livelihood. In the course of a year, we often migrate to three or four different locations. In early spring, we plant potatoes, wheat, millet, buckwheat, barley, and other local vegetables; and during the summer, we move to pastures at higher altitudes to graze yak and other animals. As a child, I spent a lot of time grazing cattle, collecting firewood, and harvesting medicinal plants.

> *In the remoteness of the Himalayas*
> *In the days of yore, the land of Tibet*
> *Observed a ban on hunting, on fishing*
> *And, during designated periods, even construction*
> *These traditions are noble*
> *For they preserve and cherish*
> *The lives of humble, helpless, defenseless creatures.*

From the beginning in the Phoksundo regions, hunting animals was restricted and forestry was controlled by local lamas. Even today you can see religious leaders at an established monastery move to stop the killing of animals and to protect the area around it from incursions of poachers, deforestation, and theft of timber. In the 1980s, most of the regions around my childhood home became protected areas under Shey Phoksundo National Park; without special permission, all forestry activities are forbidden.

Local religious leaders say that we should not destroy the ecosystem; if we destroy the ecosystem, then protective spirits of the land and water such as powerful or malicious *nagas* will harm us. Likewise, aged people say that to take care of the ecosystem is to take care of ourselves.

My people in Dolpo feel a deep connection with our ecology and ancestors. We take care of the ecosystem and have a strong concern for ecology because we have always been so dependent on it. Almost all the people have practical knowledge of biology and ecology. We know the plant history, growth patterns, and life-cycle stages of all flora and medicinal plants. We maintain full knowledge of where and when certain plants grow, flower, bear fruit, and produce seeds.

As a small child, I used to collect lots of wild mushrooms and medicinal plants with my parents, relatives, and friends. I still remember how parents and elderly people always used to carefully guide and teach the young children how to differentiate between poisonous and nonpoisonous plants, fruits, and herbs. Teaching these collecting approaches to the younger generation is considered by the elders to be a top priority. They teach a nondestructive collecting approach, an approach that aims to sustain the regeneration of plants, mushrooms, or whatever else is collected. Every effort must be made to cause no harm or destruction.

These nomads understand their geography, the space of their ecosystem, so well that when a group of elders was shown photographs of their region taken from outer space, and it was explained what the images were and how they had been obtained, within minutes they pointed out where they were at that moment, as well as numerous features and names of different bodies of water, mountains, and grasslands throughout the photos.[3]

Tibetan Plateau nomads have led a distinctive pastoral life for close to five thousand years. They have not just scraped by, either; at one point, around thirteen hundred years ago, the nomads were a vital part of the most powerful empire in Asia. While most pastoral nomads' lives around the world are shaped by lack of water, the key variables of Konchok's native ecosystem are altitude and temperature. The extremely high elevations of the plateau are too cold for growing crops, but ideal for summer livestock grazing.

The nomads and their landscape have evolved hand-in-hand with their high-altitude pastures. As Konchok notes, the plateau pastures are interrupted with those rocky mountains, dramatic lakes, and deeply carved river valleys. These diverse environments create spaces for a great diversity of species. The grasslands are high in protein and evolved with the grazing of yak, sheep, antelope, and other animals of the plateau. Depending on the time of year, altitude, wind, and water availability, different plants for grazing grow in different plateau locations —thus Konchok's migrations from place to place during the summer months with his livestock.

Like the bison or buffalo in Native American culture, the yak on the Tibetan Plateau is a mythic, ecologically vital creature (figure 5.2). Given the overall parallels between the Native American and Tibetan Plateau cultures, the strikingly analogous nature and the biologically close relationships between the bison and the yak and their respective roles in the two cultures are probably more than mere coincidence.

Yak bulls stand as high as six feet at the shoulder with three-foot-long horns. One Tibetan word for yak translates as "wealth." There is a different word for the male yak and the female. The female provides

FIGURE 5.2 Male and female yak.

the milk for the staple of butter tea. Yak hair is used for everything from tents, to clothes, to dressing one of Konchok's ankles he broke in his boyhood.

I was given the responsibility of looking after the cattle during the day. At that time, our family had more than sixty animals, which included male and female yak, cows, and *dzomo* (the offspring of a yak and a cow, or a cross between an ox and a female yak).

We used to have one black *dzomo*, and she was my favorite. She weighed around 250 kg and always walked last in the herd, just in front of me when I took them to pasture every day. She often strayed high into the mountains during the day. It is normally very difficult to chase stray animals in the evening when it is time to return home, but she was always cooperative and the easiest to return to the herd. I only needed to call her name from below and pretend that I had something delicious to give her. She would come immediately to me with big expectations; but when she found me with nothing, her expression changed and she would sadly rub her neck on my knee. I felt so bad deceiving her after a couple of times that I later started saving a little piece of heavy buckwheat bread from my lunch as a reward for her obedience.

My lifestyle in Dolpo was so simple—no competition and certainly not hectic. The routine was the same throughout the year, no matter what the season. In the morning after breakfast, I would take all the cattle for grazing a distance of two miles or more, stay the whole day with them, play with them, and then in the evening bring them back. Winter and early spring were the most challenging seasons, because early spring is the time of giving birth, and there was always the risk of attacks from jackals, foxes, snow leopards, and other predators.

I hated the winter season when I was a child, as I still do now. During winter, I always wished for heavy snow, because if that happened, the cattle stayed at home until the snow melted. Everyone would stay at home, very cozy, and the men and women would spin yak wool and weave cloth for clothing. I would have lots of free time, during which I learned how to read and write Tibetan.

Survival of my people in Phoksundo depends profoundly on the environment and climate conditions. Nature can provide recreational enjoyment and a positive impact on the inhabitants, but sometimes nature's disasters wreak havoc. I remember the time we had ten days of continuous heavy rainfall that caused devastating floods and landslides across the valley. It was one of the scariest times in my life. Many of my neighbors lost all their cattle and crops. The floods damaged many houses, and bridges were swept away. It was very painful to see my friends suffer in the grip of this disaster. That year the average harvest was barely 15 percent of normal.

Nothing exists independently of the ecosystem.

Yak are integral to the nomads' biological environment, culture, and language. Konchok notes the common saying, "Thirty people have thirty thoughts, and thirty yak have sixty horns," meaning for any single phrase, concept, or object, each person has a slightly distinct perspective, a different way to describe it according to his or her own culture and experience. The domestication of yak millennia ago probably is what made life possible at all for these nomads. Yak are integral to Tibetan festivals and rituals. There are yak dances and yak races. Nomads of the Tibetan Plateau tightly interweave nature and culture, the land and religion.

The domestication and grazing, migrating, and natural histories of yak, and also diverse species such as goats, sheep, and horses, which consume different grasses in different locations, has allowed for "sophisticated adaptive responses" by the nomads of the Tibetan Plateau, responses that take into account the available nutrients and unpredictable droughts and snowstorms of the plateau.[4]

And yet, with so much pastoralism and peace, there is always an edge of danger.

My mother was gored and killed by a yak. Yak horns are very sharp, and my mother's intestines came out. No one knew what to do. In the following days, there was a terrible infection. This kind of situation affected me very strongly.

From the moment she was hit by the yak until she took her last breath, my mother never had even a single second of relief from pain. There are several traditional doctors in my village; my father is one of them. Unfortunately, my father was not at home when my beloved mom was hit by the yak. My father did not see mom dying. Even he, with only his traditional medicines and rituals, could not have helped her.

My biggest ambition is to learn Western science, especially the life sciences, and pass this knowledge on to other monks and those in my village. My village needs a modern dispensary, modern doctors, and modern education. I have great enthusiasm for fulfilling these goals now that I am in the world of modern science, studying in India and the United States.

Konchok says his family, his village, and those living the traditional life like them on the Tibetan Plateau *are* their ecology—their every action, their lives, their deaths tightly woven into the cycles of weather, wind, water, and seasons. In his elegant writings on these nomads (with whom he has spent much time), Daniel J. Miller points out that it is not a coincidence that Konchok's seventh-century nomadic ancestors adopted Buddhism and its core values: "The landscape of the Tibetan Plateau and the Himalaya and the original folk convictions had molded the inhabitants' sensitivities in a way that was conducive to Buddhist beliefs—the idea of rebirth was one such. The belief in *karma* served to influence one's moral behavior. Even today the nomads' beliefs condition their actions."[5]

From the stories of the Tibetan Plateau nomads, it is clear that human culture and biology are part of and shape the ecosystems in which they live, and that ecosystems are part of and shape human biology and culture. In chapter 3 this two-way conversation is made evident at the level of individual organisms and populations, which are both always evolving, adapting to, and at the same time, changing the environment.

The original Bön religion of Konchok and his Himalayan ancestors evolved and was shaped by the nomadic life he describes. This life is shaped by the ecology of the Tibetan Plateau, which is partially, in turn, shaped by the nomads themselves.

Each step of the Living Staircase—from atoms, to molecules, to cells, to tissues and organs, to organ systems, to organisms, to communities, to ecosystems—both emerges from the ones below it and affects and shapes the other steps above and below it, thus inevitably feeding back to change itself. Each step of the Living Staircase includes within it, and thus is interdependent with, those steps below it.

Such interdependence might seem obvious, but the predominant narrative in the West actually views "the environment" as a separate entity from people—not too surprising given that since the Industrial Revolution and the rise of science, humans have become literally more separated from the outdoors, and less and less regularly engaged with organisms other than ourselves. This dichotomous view rears its head, for example, in the politics and discussion around issues such as natural resource use, national parks, land rights, and climate change. The environment and nature are things outside of humans, commodities, "objectified, and distinct zone[s] of activity separate from human social interactions."[6]

Such a view undoubtedly has made it less problematic for the humans who hold it to treat ecosystems more as containers of potential products than as actual systems—without careful consideration or consultation with the natives of those systems. The results? Un-mindful grazing, hunting, mining, and logging; the creation and storage of nuclear waste; and carbon poured into the atmosphere. This is what often happens following colonization—by, for example, Western Europeans in North and South America and Africa, and currently the Han Chinese in Tibet. The "settling" of the Western United States strongly parallels that of Tibet today—the forced settlement of nomadic natives, little attention given to native voices and their ecological and cultural knowledge, and the elimination of native species and damming of major rivers.

WHEN YOU HAVE TO explain your expertise in a culture and environment radically different from your own, it opens you up to new possibilities; you are more likely to see the limits of your own approaches and models. I believe a major reason dozens of scientists from around the world are attracted to and sustain an interest in teaching Tibetan Buddhist monastics is that—consciously or unconsciously—they realize the opportunity it provides for them to explore and expand the previously hidden potential, limitations, and contradictions existing *within their own science*, which they are usually too bound up in to see. This is the case with developmental biology, as we saw in chapter 2, where approaching from a Buddhist perspective the processes from fertilization to death *changed everything*. Such is also the case with ecology.

Every model has its limitations. The linear, common-cause model that dominates Western science has proven fruitful for understanding many problems and mechanisms in nature. It is often less effective, however, for addressing complex systems—such as the ecosystem of the Tibetan plateau, or the nature and causes of major depression in humans, or the collapse of coral reefs in our oceans. Tibetan Buddhism opens up a conceptual space for studying such complex systems that potentially allows us to understand and make predictions about systems for which we traditionally cannot. Intriguingly, understanding cause and effect is as central to this Buddhist space as it is to traditional Western science. Buddhism, through the concept of karma, posits that *everything across time and space* is *causally* connected, everything is interconnected and cyclical; we may not be able to explain exactly *how*, but this is the way the universe works.

Science is grounded in uncovering cause and effect; it's just that it doesn't typically consider multiple, nonlinear, nonhierarchical causes and effects. This makes predicting and intervening in systems full of such interrelated connections difficult to understand without expanding into a Buddhist space.

Just such a "Buddhist approach" to ecology was developed by a colleague of mine, Lance Gunderson, together with C. S. Holling, although Lance was as surprised to hear about the many connections I had discovered between his work and Buddhism as I was to discover that he, whom I had known in other capacities for years, was a leader in this area.

We have seen throughout this book that East-West collaborations such as the Emory-Tibet Science Initiative often lead to Western science "discovering," describing, and explaining processes and phenomena—like ecology itself—long understood as given within traditional cultures. While this could be interpreted cynically, the Dalai Lama and others see it instead as a cross-cultural reinforcement and enrichment.

I SETTLED IN across from Lance Gunderson in his comfortable office—full of natural light and gorgeously constructed from natural and recycled materials, as any good environmental sciences building should be. Lance is about as unassuming as they come. He emanates "thoughtful" and "quiet."

In exploring connections between resilience and ecology, I karmically stumbled across an intriguing concept called "panarchy," of which Lance is a lead developer. "Panarchy" is a word Gunderson borrowed from politics, originally coined in the nineteenth century to connote a system of government that would encompass all others. Lance noted that while "hierarchy" roughly translates as "sacred rules" and "anarchy" as "lack of rules," he adopted the term "panarchy" because it means "nature's rules."

Gunderson and I discussed panarchy, what it is and its parallels to how denizens of the Tibetan Plateau and Tibetan Buddhism in general think about ecosystems. He is a coeditor of the bible of the concept, *Panarchy: Understanding Transformations in Human and Natural Systems.*[7] The book has been cited over three thousand times, and not just by other ecologists, but also by psychologists, psychiatrists, economists, sociologists, educators, anthropologists, theologians, and political scientists; this is a hint that panarchy is an especially resonant idea.

What exactly is panarchy? Since the first days of the discipline of ecology in the West, ecosystems were typically thought of as hierarchical, linear levels piled on top of each other, each level of the system more or less independent, with control of the system occurring from the top down. This view is still seen in contemporary ecology texts, often symbolized with a pyramid of niches (figure 5.3). Holling and Gunderson's panarchy, in contrast, suggests that systems experience *cycles*—birth, growth and maturation, death, and renewal.

Important here, the death and renewal stages are very much integral to the cycles, to the ecosystems themselves—similar to the Buddhist idea that death and renewal are part of life. In fact, all four panarchy

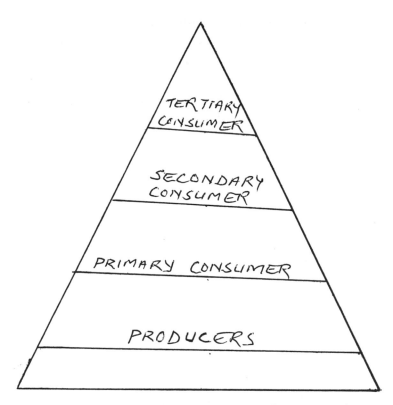

FIGURE 5.3 Traditional pyramid representation of ecosystem niches.

stages mirror the Tibetan Buddhist view of life's stages. The nature of the impermanent phenomena that are the existence of humans and other sentient beings is the same as the waves of the ocean. They begin, move, decline, disappear, and then reform. In other words, Buddhists believe *birth*, *aging*, *death*, and *rebirth* are characteristics of every impermanent phenomenon. And all phenomena, including our own lives, are impermanent.

Like the approach we took to understand the development of organisms from sperm and egg to aging adult (chapter 2), such a significant change in perspective is more than just words and abstraction, but also affects how biology is thought about and funded, the questions asked, the experiments designed, the interpretation of results, and future policies and plans.

Panarchy posits that resilience is *part of the system*. This is a different kind of resilience than "engineering resilience," in which a system returns to the same state it was before.[8] In the panarchy model, more than one stable state or "regime" exists for a system. Disturbance is not considered external to the system, but often integral to it. The very nature of resilience changes as the system changes. Systems evolve; they have an adaptive capacity to change. Gunderson suggests that rather than referring to "ecosystem collapse," such dramatic change might be thought of instead as a state of renewal, as "creative destruction" or "creative reorganization."

As forest systems go through their growth and maturation stage, using the resources of the system to accumulate leaf and wood mass, those trees become more susceptible to fire. Many tree species in such ecosystems have adapted so that they *only* release their seeds after such a fire. The fire is part of the system.

Konchok and his nomadic Tibetan Plateau society evolved as part of a unique environment. The diversity of their livestock, the diversity of that livestock's grazing needs and habits, the mobility of the nomads, the farming they do in the months when grazing is not possible are all biocultural strategies evolved over millennia—after undoubtedly extensive experimentation—so that the droughts and winter blizzards common to the region, the general risky nature of their environment, is built in and part of their "successful" existence. Gunderson emphasizes that ecosystem resilience is indeed related to the *diversity of functions* present within that system—the more the better.

Panarchy works on both fast (e.g., a particular predator-prey relationship) and slow (e.g., climate) levels or cycles. The faster, usually smaller cycles move within and often affect the larger, slower ones. Slow ones are hard to see, but are ongoing all the time. It often appears that a system suddenly shifts or collapses because of a small, fast change; but in actuality, slow cycles of the system have been changing for a long time, reaching a point near a sensitive threshold or tipping point, past which a fast change might dramatically alter the system.

We humans are fundamentally interrelated and interdependent with the ecosystem and its other inhabitants. Nothing exists independently of the ecosystem. Within our ecosystem there are two types of existence: permanent and impermanent. The concept of impermanence is the basic teaching of Buddhism. Buddhists emphasize impermanence as a principal object of meditation and use it as an instrument to help us penetrate deeply into reality. In Buddhism impermanence means that things are constantly changing or nothing remains without changes.

Change is the nature of all impermanent phenomena, and it has two levels: a subtle and a gross level. At the subtle level, we do not see the transient nature of things, but in fact changes are happening all the time. As in panarchy, this subtle level works on a large time frame and is hard to see, but goes on all the time. At the gross level of impermanence, we do notice changes in the four seasons, the aging of our bodies, and our fluctuating thoughts.

For another specific ecological example of panarchy, Gunderson turns to coral reefs. They have been recovering from hurricanes and other severe disturbances for eons, but recently some reefs appear to have suddenly collapsed, as algae have taken over the coral growth. Reef ecosystems can shift between alternate states, rather than slowly transitioning in response to changing conditions. What was going on for some time was a slow decrease in fish herbivores. This was probably because of increased amounts of nutrients, resulting from changes in land use and overfishing of first large, and then small herbivorous fish. A type of sea urchin had competed with the small fish to keep the algae in check, so with the decrease in small fish, the urchins increased in number; however, after a pathogen wiped out the urchins, the algae took over and the whole system shifted.[9] Several different factors in a web of several different cycles were crucial; the system collapsed because of what seems a relatively small change, but when all these cycles happened to interact in a particular way at a particular time, the system shifted. If we can understand how these factors are related and which ones might predict a major shift in the system, we could develop better strategies for sustainable management of the resilience of these systems.

To further illustrate the potential impact of thinking from a "panarchy perspective" rather than a hierarchical, linear one, let's take an example of an "ecosystem" that's a little easier to wrap our heads around: our own human resilience and, specifically, major depression. How might panarchy theory relate to or change the approach to the mental and physical resilience of people?

Resilience is a major theme of this book—resilience of individual Tibetans such as Tenzin and Konchok, resilience of the Tibetan culture, of the Tibetan Plateau. It is implicit in discussions of response to stress, meditation, compassion, and the impact and interaction of genetics, epigenetics, and the environment. Studying human resilience from the perspective of panarchy leads to an expanded, more nuanced definition of this resilience, in much the same way it has for ecosystem resilience. Human resilience, too, is much more than simply returning to the same previous state after a major disturbance. Resonating with Buddhist ideas of self and karma, psychiatrist Rachel Yehuda calls resilience "a process of moving forward and not turning back . . . a reintegration of self that includes a conscious effort to move forward in an insightful, integrated, positive manner as a result of lessons learned from an adverse experience." If it's just a trait, genes are important, but Yehuda emphasizes that resilience is more a *process*; so in this case, it's how environment affects those genes, the nature of social and therapeutic support, that is important.[10]

The psychologist Ann Masten refers, as we noted, to human resilience as "ordinary magic," that is, most people are remarkably resilient; it is part of the "adaptive capacity" of the human species, a vital characteristic of what makes humans human, just like the trees for which fire is part of their life cycle.[11] Only a small minority of soldiers suffer post-traumatic stress following deployment, even a lower percentage after a second deployment.[12] As Kerry Ressler points out in chapter 4, post-traumatic stress disorder is *even uncommon in those who are genetically susceptible to it* in the large inner-city Atlanta population he has been studying.[13]

Now to our concrete example of panarchy in major depression. Despite the fact that 17 percent of all people on earth will suffer some form of major depression in their lives, we have little understanding of how depression works or how to predict it and, thus, of how to treat it effectively. Very few interventions work well, and then for only a very few people. Rather than consider, then, that major depression results from a common cause, instead imagine, like coral reef collapse, it's the result of many interacting cycles and components—emotions, awareness, behavior. So, Marten Scheffer reasons, one may become depressed in a linear, causal chain from stress, leading to negative emotions, sleep problems, and feeling depressed or unable to feel pleasure (the thicker arrows to the right across the middle of figure 5.4). But

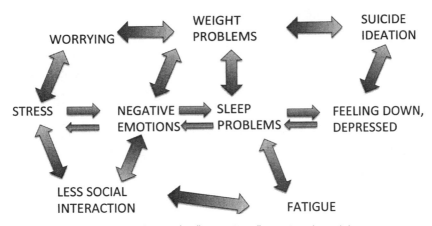

FIGURE 5.4 A complex "ecosystem," or network model, of cause and effect for human depression.

now we add the interacting networks of panarchy thinking, only just approximated in figure 5.4. First, many other processes and cycles are ongoing simultaneously, which might also lead to major depression or feed into other processes that do (a limited portion of the network of such possible links is indicated by the additional symptoms and arrows in figure 5.4). Then, even within the one linear cause-and-effect process we started with, the arrows actually move in two directions (indicated by the thinner arrows pointing back to the left)—lack of sleep might make one feel *more* negative emotions, and these might make one feel more stressed. And other reinforcing phenomena may be occurring, other cycling and feedback, interaction between symptoms—such as worrying, leading to feeling down, cycling back to create more worrying and feeling down, and then this leading to less social interaction, which might cycle back to feeling more down. Existing data on major depression are consistent with such networks of interaction, containing nests of causal relations, rather than having a common-cause, linear explanation.[14] Note that the phenomena in figure 5.4 are all causes or effects of the conversation between genes and the environment we have been discussing throughout the book.

Complex systems of this nature, like coral reefs, can have alternate stable states or regimes. These can shift from one to the other because of enormous external shock, stepwise change, or a very slight change in the system, such as the urchin pathogen, resulting in a sudden and huge change. We focus on the latter and the tipping point just prior to the huge change. If we can figure out what the specific changes are when the tipping point is being approached, the implications are dramatic: in the case of major depression, for example, the potential to prevent it or to spur the process of emerging from it. Because of the massive complexity of systems like human resilience, mathematicians have tried to identify tipping-point indicators that are true *regardless of the variables measured and regardless of underlying mechanisms*. Remarkably, at least one such phenomenon—"critical slowing"—is observed that fulfills this criterion for complex system tipping points, whether they be in major depression, coral reefs, lake ecosystems, or financial markets. This is absolutely mind blowing and resonates with the Buddhist claim that everything is connected to everything else.[15]

Critical slowing refers to slow recovery from small changes near a tipping point. As these systems near their tipping points, the time needed to recover from a disturbance becomes longer and the systems become more correlated to their past, leading to an increase in what is called autocorrelation—the correlation of a value with its own past and future values. Autocorrelation increases near a tipping point, and systems are more vulnerable to dramatic shifts at these times.

By using a self-reporting survey of mood across several days (focused on four emotions—cheerfulness, contentment, sadness, and anxiety—thought to together give a simple but effective measure of mood state) on a cell-phone app, Scheffer and colleagues studied a general population sample that varied in the eventual development of depression—as determined in a follow-up study. In the 13.5 percent of this population that became clinically depressed, critical slowing of the mood variables measured did indeed occur just prior to the tipping point into depression. Additionally, in an analysis of similar data from a population of already depressed patients, critical slowing was also seen in these variables in that small subpopulation of those patients who recovered from depression. While emphasizing that their data in no way imply that *all* cases of major depression in everyone will follow the tipping-point model, Scheffer and his colleagues do suggest their approach might be used to predict major mood shifts and, together with therapies, provide a new avenue for preventing or improving symptoms of depressive episodes—even without knowing the mechanisms of what is causing the symptoms.[16]

THE ENLIGHTENED GENE

In at what first appears a gigantic leap from this work, Scheffer has also collaborated with microbiologists to apply panarchy concepts to human microbiomes. As we have seen, the organisms that compose human microbiomes are perhaps as much a part of humans as our own cells—a striking example of interdependence—affecting our mental and physical health.

Scheffer and his collaborators, by analyzing one thousand human microbiomes, demonstrate that the diversity and nature of these microbiome populations have two stable states, each containing particular distinctive species. The researchers propose that these distinctive species are "tipping elements" of two different states between which microbiomes can switch. In a given individual, these species are either prevalent or virtually absent (and do not shift in response to short-term changes in diet) and might be used to better understand and manage human microbiomes and their relationship to health. For example, different tipping elements correlate with the age of individuals, while others correlate with levels of obesity. The suggestion is that the resilience of a given stable overall state is related primarily to just a few key species of bacteria that can be followed as harbingers of change.[17] It would be intriguing to look for "tipping element" species that correlate to the autism discussed as related to the microbiome in chapter 6.

THE CYCLE OF BIRTH, growth, destruction, and reorganization occurs continually at every level, and therefore, change, "experimentation," and adaptive capacity are built into and occur at every level of the system. Panarchy might actually be, then, a kind of reenvisioning of the Living Staircase as an interacting network, rather than a hierarchy of steps.

The origins of panarchy-like thinking can be traced back to analyses at the molecular level and concepts discussed by physicist P. W. Anderson in his famous paper, "More Is Different: Broken Symmetry and the Nature of the Hierarchical Structure of Science."[18] Here the challenge was again to explain sudden shifts between two states—in this case the physical states (an example of Anderson's "broken symmetries") among the same group of molecules. A common example is the state of water molecules as water changes, upon boiling, from liquid to vapor.

The study of the magnetic state of some materials is another striking case; when heated above a specific point, called by definition the Curie temperature, ferromagnets are only weakly magnetic. When cooled, they remain at this level of magnetism, but only until *exactly* the Curie temperature, at which point they become strongly magnetic and remain so as they continue cooling and indefinitely thereafter. As with the other complex systems we have discussed, several mathematical properties of ferromagnets "behave oddly" near the Curie temperature transition point. For example, the susceptibility of the system to a small applied magnetic field is very low well below or well above the Curie temperature; but nearer to it, the system becomes increasingly susceptible, and infinitely so at the Curie temperature. Sound familiar?

In sum, then: as these complex systems get closer to their tipping points, it takes them much longer to recover from even very small disturbances than the same-size disturbances farther away from the tipping points. This means the system becomes more correlated to its past the closer it is to its tipping point. Because of this, small disturbances accumulate, leading to an increase in the fluctuations of variables near the tipping point. All of these characteristics can be translated into measurable mathematical processes. And all of these systems behave similarly *independent of which variables are being measured and independent of the mechanisms responsible for the change in those variables*. These characteristics appear to be *inherent to the nature* of these complex dynamic systems. This both resonates with Buddhism and provides a framework within which scientists can work to study these systems. The goal would be if not to *prevent* tipping points, then to predict, prepare for, and better understand them. And ultimately, to decrease suffering.

The panarchic nature of complex systems sounds very much like karma, and is reminiscent also of the biological concept of emergent properties explored in our discussions of cell sentience in chapter 1 and again later in discussions of consciousness in the next chapter. The concept of emergent properties is often called on in science to explain phenomena that are "more than the sum of their parts"—as we do in chapter 1 in explaining how hydrogen and oxygen on their own have radically different properties than the water that they can combine to make. But what if "emergent properties" has filled in for biologists as a fancy way to avoid saying "properties we can't really explain with our current understanding of things (i.e., linear, common-cause models)"? Perhaps the more Buddhist-panarchy way of thinking could provide a nonlinear, yet scientific, multiple cause-and-effect network of pathways to help unravel the true underlying nature of many emergent properties—a complex system of recurring cycles that, when interacting in particular ways at particular times, can result in what we see in nature.

> *The interdependent nature*
> *Of the external environment*
> *And people's inward nature*
> *Described in tantras*
> *Works on medicine, and astronomy*
> *Has verily been vindicated*
> *By our present experience.*

I am reminded of a teachable moment with Konchok and his peers in Dharamsala early on in our project. Often such moments reveal more about the teacher than the students. We were discussing connections between disease and the presence of bacteria. This led to our describing the distinction between correlation and causation, how scientists differentiate and address them experimentally. The monks were struggling. Suddenly, it became clear why: if, in one's worldview, *everything* is causally connected and interdependent, there is really no such thing as mere correlation.

Is it possible to predict *when* ecosystems, such as those of the Tibetan Plateau, are near their tipping points? Can we identify variables to measure in such exceedingly complex systems that would allow the prevention of, or at least the timely preparation for, such dramatic flips? What will the new regimes look like? Such is the nature of the questions that emerge from this different way of viewing ecosystems.

> *Our obdurate egocentricity*
> *Ingrained in our minds*
> *Since beginningless time*
> *Contaminates, defiles, and pollutes*
> *The environment*
> *Created by the common karma*
> *Of all sentient beings.*

In the 2000s, communist political activities greatly impacted the ecosystem of Phoksundo. Many species have gone extinct because of changes in climate, atmosphere, the decline of ecological resilience, and destruction of the environment. Many rivers originate in Dolpo, such as the Bheri River, which flows through many valleys before joining the Karnali River and entering the plains.

One of the negative effects on the Bheri ecosystem is that over the last two decades, many of the water species have gone extinct because of the inflows of pollution and sediments. It was difficult for these species to adapt and tolerate it, and their physiology could not cope with these situations. Traditionally, most of the mountains remain snowy all year round; but recently, because of global warming, the snow has been melting more, and sometimes the mountains have no snow.

Here, striking again in its relation to Tibetan Buddhism, *memory* becomes very important, as if the karma of the last state of the system affects the possibilities for the new life of the system.

The shuffling and reorganization of the system during renewal is largely dependent on what was present in the previous state—the evolutionary memory (say, the pines only releasing their seeds after a fire), the resource memory (the actual physical materials still available), and the genetic and epigenetic memory (that is, the experiences) of the remaining species.

This reshuffling also naturally allows for new members to join the community, seeing an opening, as it were, for new opportunities and experimentation. When is an "invasive species" no longer considered invasive, but instead native? What measurable factors affect the Tibetan Plateau, source of much of Asia's water? Intuitively, global warming would be predicted to have a greater impact on the Plateau, given both its altitude and its relative proximity to the production centers of China. That nation releases gigatons of carbon dioxide from burning coal into the atmosphere.

Carbon dioxide is called a "greenhouse gas" because it traps heat and leads to the warming of the earth. Indeed, a recent report from the Chinese Academy of Sciences and the government of Tibet states that the temperature on the Tibetan Plateau has risen at twice the already problematic global average rate since 1960.[19]

This heat is shrinking the glaciers quickly. At the same time, the increase in temperatures is melting the vast areas of plateau permafrost, an important carbon sink. This, together with the fact that precipitation has also increased 12 percent in the same period, has led to an increase in the number and size of glacial lakes.

> Perennial Snow Mountains resplendent in their glory
> Bow down and melt into water
> The majestic oceans lose their ageless equilibrium
> And inundate islands.

The potential results of these ecosystem changes, some of which are already evident, are staggering: increased flooding (with the accompanying increase in waterborne diseases and decrease in access to clean water), alterations in the monsoon patterns Asia relies on for agriculture, damage to natural fisheries, longer droughts, lack of water for the production of hydropower (for China's thousands of new dams, for example).

During my visit to Dolpo in 2015, a few months before the massive earthquake there, I was reminded how the people of this area are so calm, compassionate, and hardworking. I never saw expressions of depression, frustration, or desperation on their faces. My people astonished me, as they seemed to have no anxiety, no depression—even with their very difficult lifestyle.

Males and females, children and elderly, all family members work hard in the fields and in the home. The children face so many risk factors while they are growing up. Some children who grow up with a number of risk factors do not do well later in life, but I noticed that most of the children in Dolpo who have been through a hard life earlier do better in later life.

One example is my sister. On this visit, I saw her for the first time in eight years, and she told me her story. She faced a series of tough challenges in her early life. When my mom passed away, my sister was the only female in our house. She was only fourteen, yet she took on all the female daily chores of the house. Because of strong cultural influences, it was her duty to do the chores even though she was so young.

A few years later she got married, but in her marriage she suffered continual physical abuse, often being knocked unconscious; and she left her husband and returned to our home. Now my sister lives by herself in a place that is two days' walking distance from our family. She runs her own small hotel; she has become stronger and more aware. Twice a month she takes the long walk to visit my dad and check on his situation. She has continued this despite the earthquake, which flattened my dad's house and destroyed so much. My dad was injured, but not badly, and they have rebuilt his home.

My sister blames no one for her situation. She says it is karma, a way our culture uses for coping. If this same incident had happened to a young girl in another culture, she might have committed suicide. Instead, my sister is strong and confident.

Education, too, is an ecosystem. Again, such a claim is not trivial: if education is a dynamic web of interrelated phenomena, rather than an information-in/information-out assembly line, then how we develop, engage in, intervene in, and evaluate it—all the most vital elements of relationship with the system—also change. Grant Lichtman, a writer and researcher in K–12 education, says: "This is not a metaphor; ecosystems are governed by a set of laws that are very different from those that govern an efficient assembly line. Education should not *act like* an ecosystem; it must *be one*."[20]

This is not a new idea, of course. Even though the word "ecosystem" had yet to be coined, John Dewey, a prominent American thinker in education, was proposing similar ideas in the early 1900s. In his influential *Experience and Education*, Dewey provides a theoretical foundation for the nature of education that still rings true, both in the American classroom and with Tibetan Buddhism. He sets up a dichotomy between "traditional education"—the structured, assembly-line, information-based approach to teaching—and "progressive education," in which the classroom is given over to the students; their stories, beliefs, and experience drive their learning.[21]

Dewey says an ideal education integrates these two philosophies. For him, educators are facilitators who must understand the critical link between teaching, learning, and experience: "To know the meaning of empiricism we need to understand what experience is."[22] Fundamental to learning are *continuity* and *interaction*.

Educators should give students a continuum of quality experiences from which they can draw; experiences, whether good or bad, build and accumulate and impact experiences to come. Effective educators select and shape these experiences for learners, while seeking to understand students' past histories. Past and present experiences interact to help create the future. For Dewey, educational experiences are authentic when the internal experience of the learner is at least as, if not more, important than the teaching itself. Traditional education can be too focused on objective, external goals.[23]

Education is an interaction among learners, objects, and teachers; and educators need to pay special attention to the environment of learners, including both their social and physical surroundings and "whatever conditions (objects or people) interact with an individual's internal personal needs, desires, capacities, and purposes that create the resulting experience,"[24] that is, their ecosystem. Like the Dalai Lama, Dewey emphasizes the importance of the community, how the educator must be aware and part of the classroom and greater societal community, and how enabling and strengthening community improves learning.

With his karma-like ideas about the continuity of experience and the importance of past experiences affecting the present, and his focus on the interaction between the external and the internal and the greater ecosystem, Dewey appears to be working right between biology and Buddhism in shaping his educational philosophy.

Urie Bronfenbrenner, a developmental psychologist, and others went on to extend these ideas of educational context and community; they actually adopt ecological language in their discussions of education and how best to study it. Bronfenbrenner's "ecological structure of the educational environment" includes the microsystem (the immediate setting of the learner); the mesosystem (the interrelations among the immediate setting and the other groups in the learner's life, such as peers, family, religious groups, and camps); the exosystem (the formal and informal social structures, such as neighborhood, work, government, mass media, social, and transportation networks); and macrosystems (the greater economic, education, legal, and political structures). He shows how understanding education in this way, as a complex system, is crucial to how it is developed, studied, and evaluated.[25]

LANCE GUNDERSON also points out a parallel between Dewey's philosophy and his colleague C. S. Holling's ideas on resource management, which Holling calls "adaptive management," and bases it on panarchy and resilience theory. Adaptive management strategies are being used around the world. The approach considers policies as hypotheses for, rather than as solutions to, resource problems; actions do not implement policies, as much as they test or evaluate them. For both Dewey and Holling, actions need to be evaluated and changed; we need to learn by doing.

The Emory-Tibet Science Initiative forces us to look carefully at how we teach science and to rethink the way we teach "back home." Most of this rethinking integrates well into a teaching-as-ecosystem model. When science educators teach evolution without a social, political, and religious context, the implication is that science knowledge exists in a vacuum and does not effectively integrate and converse with the greater ecosystem of knowledge and society.

Knowledge in a richer context is knowledge better experienced. When students can "design their own syllabus" based on their own lives and experiences, what they are most interested in, the questions that draw them in the most (à la Dewey), they own that knowledge and learn it better. When students, their lives, and their personal stories are explicitly made part of the living and learning ecosystems they experience, they learn better and are more likely to become contributing scientists or more knowledgeable citizens. Recall the students in the ethics, religion, and biology seminar that they took together with their course on cell biology (discussed in chapter 4). When they explored their own stories, beliefs, religions, and ethics related to and about the science *as they learned the science*, they learned it in a rich ecosystem context. As one student recalled: "Many times we tend to think that each aspect of life doesn't really affect us at the cellular level. But the way . . . [the course] content guided us through social, religious, [and] behavioral aspects provoked us to make the links between each other and to cells which ultimately make up who we are."[26]

The very process of translation that the Emory-Tibet Science Initiative requires is an exercise in ecosystem management. What are the different languages employed? How are they best translated across national, religious, and generational cultures and systems? These are teaching-ecosystem questions. "Do you mean 'cell minds' or 'cell bodies'?" Geshe Dadul asked, when translating the cell biology textbook into Tibetan. "When we [Tibetans] refer to sentient beings, we speak either in terms of their bodies or their minds; which do you mean here?" The question neatly summarizes the entire challenging web that is our project—are single cells sentient beings? What are sentient beings anyway? How do they undergo birth, death, and rebirth? Why don't Western scientists think of cells as having minds? Or as having bodies for that matter? Once we translate it, do we explain why we translated it that way?

Again: all teaching is translation. And that translation is interdependent with the many components of the dynamic complex ecosystem that is education.

Irony abounds here, because the concept of an ecosystem itself is a product of scientific thinking and because a major reason the Dalai Lama catalyzed the Emory-Tibet Science Initiative in the first place is his conviction that education is the answer—to cross-cultural progress and interaction, to the integration of modern science into the monks' world, and vice versa. And yet, it is science education that is perhaps most guilty of *not* approaching education as an ecosystem, but more as an assembly line. Even today with all the research, ideas, and anecdotes about education—much of which we have discussed in this book—thousands of college students still "learn" science in massive lecture halls, one way in, one way out.

There are some good signs. The word "community" does appear fifty-seven times in the sixty-one pages of one of the latest big national reports on how to improve American science education—a report with the hopeful title "Vision and Change in Undergraduate Biology Education: A Call to Action," published by the American Association for the Advancement of Science and the National Science Foundation. "Community" is used in relationship to a national and campuswide community of scholars; development of this "academic life sciences community"; and a need for change by and for this community. The report calls for including and connecting students in this conversation as a vital part of their education in the classroom, field, lab, and beyond.[27]

The report does raise some of the teaching ideas emphasized in this book, a bit of Dewey's progressive education. Science educators should integrate facts into concepts, emphasize process, provide less information in more depth, engage students as active participants, diversify pedagogy, employ diverse and continuous feedback, and provide rich experiences and group work in the classroom. While compassion and empathy are not mentioned, and passion only in terms of what scientists should demonstrate for their discipline, "reflection" *is* mentioned twice—once generically and once as in thinking about knowledge.[28]

Advocates of contemplative and transformative education propose one more step—a big one. As Arthur Zajonc, a physicist, puts it: "The solutions to . . . economic inequality (among other problems) will ultimately not be found through more information or better foreign aid programs, but only here at the level where information marries with values to become meaning. Human action flows from this course, not from data alone."[29]

As we hear directly from the Dalai Lama in the following chapters, secularized methods of reflection, such as meditation and other contemplative practices, might be an effective way of merging data with meaning and ethics.

> *Even in this age of science*
> *Of technology*
> *Trees provide us shelter*
> *The chairs we sit in*
> *The beds we lie on*
> *When the heart is ablaze*
> *With the fire of anger*
> *Fueled by wrangling*
> *Trees bring refreshing, welcome coolness.*

I took my *geshe* exams in March 2014. It is perhaps the greatest achievement of my life. It took me seventeen years to complete these studies of Bön philosophy, grammar, astrology, astronomy, and traditional art—the equivalent of a PhD in religious studies.

It was so wonderful to have my family, especially my father and elder brother whom I had not seen in eight years, with me during my *geshe* exams and all the precious ceremonies that accompany this experience. The most touching thing for me was the fact that, although my father was seventy-seven years old, he carried his backpack and walked for days, crossing many rivers and mountains just to get to the airport so he could attend my *geshe* ceremony. His great joy and enthusiastic pride for my accomplishment was a very emotional experience for me to witness. I feel very fortunate to have been able to share with him all the stories about my journey so far.

After the *geshe* ceremonies were over, I went to the Gaden Monastery in the south of India. There I worked alongside other monks who studied at Emory and professors from the United States to implement the first science course ever at that monastery. I acted as teacher assistant and translator and participated as well as a student. Translating for the monks and hearing their views about what was being taught was an enormously enriching opportunity. Acting as an interpreter was especially demanding and fulfilling. It was transformative.

CHAPTER SIX

Are Humans Inherently Good?

We are all born free of religion, but none of us
are born free of the need for compassion.
THE DALAI LAMA

While we have this body, and especially this amazing human
brain, I think every minute is something precious.
THE DALAI LAMA

hether one is a monk or not, the idea that humans are *electrical beings* gives pause. There are these billions of cells inside you, scientists tell us. They produce electricity spontaneously or upon receiving signals from other such cells. And guess what? We can measure the electrical activity of these cells and relate it to specific movements, thoughts, ideas, and even feelings such as empathy. We can even measure such activity, at least in monkeys, who share much of human brain anatomy, from a *single one* of these electrical cells.

Westerners take such claims in stride nowadays. But even jaded neuroscientists sat up and took notice in the early 1990s when a group of Italian scientists serendipitously found individual electrical cells—neurons—in macaques that produce electrical activity both when the monkeys pick up a piece of fruit and *when they watch another monkey* perform the same action.[1]

The so-called mirror neurons are not only a striking scientific finding—after all, if empathy is "a complex process that involves sharing an emotional state with another organism,"[2] it is difficult to imagine a better potential mechanism for it than through such cells—but their discovery resonated broadly within the Western science narrative. Now empathy, a previously relatively vague and undefined emotion from the neuroscientific angle, could potentially be analyzed, explained, and measured.

Soon mirror neurons, although to this day not yet specifically identified in humans at the cellular level, were proposed as key players in, among other complex processes, imitation, compassion, language, and diseases such as autism—and that is just in the peer-reviewed scientific articles (more than fifteen hundred of them as of this writing). By 2000 the respected neuroscientist V. S. Ramachandran was predicting that "mirror neurons will do for psychology what DNA did for biology: they will provide a unifying framework and help explain a host of mental abilities that have hitherto remained mysterious and inaccessible to experiments."[3]

What is the evidence for mirror neurons? If they do indeed exist in humans, would they contribute to an understanding of empathy and compassion, and if so, do they provide a way into a joint science/Buddhism exploration of such capacities? Why did these little electrical cells get so famous so quickly, and what does this say about science in a cultural context? These are the questions that drive this chapter.

I learned that the human body is like a complex machine, an electrical being, and that we are made of many little parts like cells that communicate or work by themselves or with others via electrical or chemical signals to keep us alive and regulate motion, thought, and many other things.

I also learned about neurotransmitters, which help transmit nerve impulses from one nerve cell to the other. Neurotransmitters are tremendously vital to us because we could not function without them. We would die without them.

To me this was not surprising. Neurotransmitters are like *rlung* (vital energy) in Tibetan medicine. *Rlung* [pronounced "loong"] is the wind or breath that helps to regulate the human body, to create motion and thought. Without *rlung* we would not move, and we would die.

There are many types of neurotransmitters, and actually the functions of the body depend on those neurotransmitters. Likewise, Tibetan texts mention that there are five types of wind—life-sustaining wind, ascending wind, pervading wind, firelike wind, and descending wind—and all the functions of the body depend on these winds. For example, life-sustaining wind resides at the center of the body and affects mind/body interactions, and ascending wind affects speech, awareness, and memory. These five winds work together just like the different combinations of neurotransmitters—serotonin, dopamine, and glutamate—work in balance to affect different mental and physical actions and states. And disease results from an imbalance of the winds or neurotransmitters.[4]

Back in Atlanta, my family has a tradition just before digging into our Thanksgiving feast. Each person around the table says something he or she is especially thankful for that year. One year the table extended to fill every inch of the dining room—the air thick with intermingling smells of sweet potatoes, turkey, brussels sprouts, casseroles, and the promise of my mom's pecan and apple pies.

We invited over Ngawang, one of the monks who was part of our pilot project in Dharamsala and was now studying at Emory, for his first Thanksgiving. Like many of the monks, Ngawang has a presence, a kind of serenity and way of listening that make him seem much larger than his slight frame. He has since gone on to become a leader in science education among Tibetan monastics in exile and has already set up a science center at his monastery.

At the table, it is Ngawang's turn to give thanks. "It is my wish," he says, "that one day Tibetans and Chinese will sit down at the same Thanksgiving table and exchange blessings as we are now."

Ngawang's words take me back to a cramped apartment in Dharamsala. Several students and I had gone to visit a diviner. This monk used several dice and complex formulas to provide advice to anyone who stopped by. He was among the happiest people I had ever met; and as he rolled out the fates of one after another local Tibetan who tucked their heads in his doorway (Should I go to the traditional doctor or the Western doctor? On what day should my daughter get married?), the diviner spoke to us of his own fate, his eyes incongruously twinkling. Before escaping to India, he was imprisoned and tortured in Tibet by the Chinese police. The police could not help it, he says; this was their job that they must keep to support their families. Other monks talk of similar experiences and, after their release, of seeing their torturers in the street and inviting them to dinner.

Compassion and empathy undergird the noble truths of Buddhism, but where does the kind of compassion and empathy exhibited by Ngawang and the diviner come from? Are such astounding capacities of forgiveness and understanding unique to monks? To humans? Are they outside the realm of science, or do these monks have more (or more excitable) mirror neurons or other specialized cells? Can such traits be nurtured and even measured like other abilities? These are profound questions extending into that loamy space between biology and Buddhism. How do we most effectively sow and grow new ideas in this space?

LET'S START with flowers—those often-flamboyant sex organs of plants both sown and wild. If we can understand flowers, these common physical elements of our everyday experience, as an emergent property (keeping in mind the limits of this concept plumbed in the last chapter), perhaps that will help us understand the harder-to-encapsulate ideas of empathy or consciousness as analogous emergent properties.

Years after we used flowers to teach sex to the monks and nuns, I found this in the Dalai Lama's *An Open Heart: Practicing Compassion in Everyday Life*: "If we analyze or dissect a flower, looking for the flower among its parts, we shall not find it . . . And yet, we cannot deny the existence of flowers and of their sweet scent . . . Within the context of a single phenomenon like the flower, its parts—the petals, stamen, and pistil—and our thought recognizing or naming the flower are mutually dependent. One cannot exist without the other . . . Therefore, when analyzing or searching for a flower among its parts, you will not find it. And yet the perception of the flower exists only in relation to the parts that make it up."[5] This insight reaches into the heart of our project: how do scientists and monastics study and appreciate the parts while not forgetting the wholes that emerge from them?

We were initially hesitant to teach celibate monks and nuns about sex and reproduction, but my friend Lobsang, the former monk and leader of our project, insisted we do so and assured us it would be no problem.

The Dalai Lama has been very clear that he would like both monks and their monasteries and nuns and their convents to be very involved in our project. When our collaboration with the monastics began, the few nuns in the class stuck together and sat quietly in the back of the room. Over the years, this changed, and monks and nuns became friends, sat together, and exchanged ideas.

I usually am a quiet and shy person. Because of that it is hard to get acquainted with people, especially nuns. It was my first time learning with nuns. It was good having class with them. They were not as I thought. I thought they might be feeling uncomfortable, shy, and would not be frank. Sometimes in the Tibetan community women are taught to be shy, and naturally they feel uncomfortable when there are large numbers of men around them discussing topics like sex and reproduction. But the nuns were very confident, and when I saw that the nuns were so open, it made it easy for me to talk and discuss with them.

The Buddhist monastic rules, known as the Vinaya, specifically prohibit sex. Literally, they say, monks or nuns are not allowed to even think about sex. Even if you think about sex, you are not allowed to talk about it. Even if you talk about sex, you are not allowed to engage in sexual intercourse. You are totally prohibited from thinking of sex, talking about sex, and having sex. Therefore, as a monk at the beginning of my studies of biology, I felt slightly uncomfortable; but, in fact, sex is one of the most important parts of biology. It is one aspect of knowledge that everyone should learn.

Buddhism says, one must be fully aware of everything to become fully enlightened. Thus, one must not limit one's knowledge to Buddhism, but must include all areas of knowledge.

As always when we leave the classroom and go into the lab or outdoors, the monastics become especially excited. We set up lab stations in the courtyard outside, complete with flowers found growing nearby or purchased from a local florist. The monastics have dissecting instruments, magnifying glasses, microscopes, paper, pencil, and each other. We review the parts of a flower; discuss how most flowering species have male and female parts in the same flower and others, in separate plants; how there are many nonflowering plants that reproduce using seeds or spores; and how the basic elements of reproduction and development found in many plants are very similar to those in humans. The monastics often find evolutionary parallels and commonalities like this between plants and animals odd, since traditionally they consider

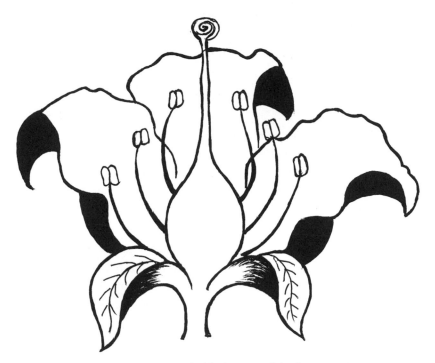

FIGURE 6.1 Konchok's drawing of the flower
he dissected in Dharamsala.

plants to be entirely different, nonsentient organisms, distinct from humans and other animals.

The monks and nuns form groups and begin to study their flowers, dissect and draw them, and find pollen grains and examine them under the microscope. Many of their drawings are startlingly complex and accurate (see Konchok's drawing of a flower in figure 6.1).

As the Dalai Lama predicts, we do not find the flower among its parts; the flower and its sweetness instead appear as emergent properties of those parts. We consider how to appreciate and more deeply understand both the parts' properties and the new properties that emerge from their interaction along the Living Staircase. This kind of teaching and learning has the feel of contemplation and illuminates how our project serves as its own laboratory of science, education, and religion, for exploring such emergence.

It's more than merely the fact that we are teaching men and women devoted to a life of spiritual learning, cloaked in robes, their heads shaved to a shine; it's also the meticulous attention, near reverence, that these monks and nuns give to their assignment of dissecting and drawing the flowers; their excitement as they call friends over to see thorny lunarlike pollen grains under the microscope; their realization that plants are indeed as alive as they are.

When I started learning biology, I was really fascinated by the scientific way of studying, describing, and explaining natural phenomena through observation and experimentation. During that time, I realized that Western science has had remarkable success in explaining, describing, and observing the functioning of the natural physical world much more than Buddhism has.

I had studied Buddhism for fifteen years, but I did not know the details of plant reproduction, and I sometimes wondered about it. In Buddhism, normally we do not focus on the different parts of a flower and their functions required for survival. It was so interesting to learn these things.

It was a revelation to me when I realized that a good way to learn about the life cycle of a plant or other living things is by sorting out parts, labeling them, analyzing them, and learning useful information every step of the way. It is very important to discover what each part does and how it contributes to the life cycle of a living thing. In Buddhism, to learn about the natural state of any phenomenon, we do not use dissection or otherwise sort out the parts of the object.

According to Buddhist epistemology, if you want to study the reality of a phenomenon, you cannot find it through sorting it out and taking it apart. Buddhist philosophy says that phenomena do not truly exist as they physically appear to be; this is merely appearance. Perhaps the Buddhist view is closer to molecular physics than biology, because, according to physics, when you get down to the molecular level of phenomena, everything is filled with empty space. So what we see on one level is totally different at another level of perception.

In Buddhism it is difficult to find the true essence of a living being in one designated part of the object. Take the example of a flower. In order to analyze a flower as it appears, you would find it difficult to point out the essence of the flower if you take it apart.

The question becomes, where is this essence located? Even at the objective level, once you have taken the flower apart, it becomes difficult for most ordinary people to recognize the parts and identify the flower. One part of the flower would not designate a flower. The topside of the flower is not a flower; the bottom side of the flower is not a flower; the pollen is not a flower; the sepal is not a flower; the carpel is not a flower; and the petal is certainly not a flower. Only when each and every part is together can we then call it a flower.

While I was learning that plants reproduce in the same basic biological way humans do, I was fascinated and amazed. This enriched my insight, and it has slightly changed my thoughts about these things. Many times I thought there were no reasons for not accepting that plants are beings like humans. But when one looks more carefully there is one major difference on a more emotional and psychological level. Buddhists believe that even though plants have the ability to grow, to move under their own power, to reproduce, to process energy from food, and to respond to the environment, they are not sentient beings like us. This is because plants probably do not have the capacity to experience pain and pleasure, nor do they have preferences, desires, and cognitive activities. How might scientists test for the presence of desires and feelings in a plant? Obviously, it is not possible to answer this question by looking through the microscope at either human beings or plants.

In the same way the flower is an emergent property of petals, stamens, and pistils, compassion and empathy are emergent properties of brain and mind (and perhaps, of cells such as mirror neurons).

Why does a biology of empathy sound not quite right to the typical Western scientist? Let's consider this question, this tension point and its sources, as a door into developing just such a discipline of inquiry.

First, the scientific worldview posits that any behavior, thought, feeling, or idea ultimately can be explained at the level of, and indeed arises from, molecules, genes, and cells. Second, scientific innovation, research, and the interpretation of that research are very much shaped by societal context; and the inevitable result is a focus on certain paradigms at the expense of others. Third, scientists and Westerners in general (ironically in part *because of* science and technology) have come to expect from science linear, quick, and easy explanations of how the world works.

Given that scientists are aware of these obstacles and, similarly, Buddhists are aware of their own equally challenging obstacles, how might such research, in collaboration with monastics, result in new kinds of flowers and move this knowledge into new realms and applications in human interaction, education, and relief of suffering?

WHETHER COMPLEX EMOTIONS such as empathy and compassion can emerge from mirror neurons or any other cells is a hard enough problem, but neuroscientists call consciousness *the* hard problem. Consciousness is perhaps the most astonishing of emergent properties. We began this exploration in chapter 1, asking, are bacteria sentient? That is, do they have consciousness or mind?

Neuroscientists make the strong claim that consciousness emerges from the brain and its vast collection and collaboration of neurons and their supporting cells. In the fast-expanding field of neuroscience and certainly in this book, we have begun to see how studying the brain can reveal and enlighten. It's not surprising that neuroscience—and its meteoric rise in technologies and resulting approaches for analyzing the brain—captured the imagination of the Dalai Lama and was a major catalyst for his vision of the creation of our project.

The very idea that consciousness is an emergent property of cells and molecules, though, is exceedingly science-centric and challenges a strong lineage of *Western* thought going back at least to the Greeks and much later developed by the Romantic movement. Remember in chapter 5 where we point out "emergent property" may just be scientists' authoritative-sounding way of saying "this happens in some way we don't understand at all." Others claim there may indeed be measurable units of consciousness composed, say, of a particular structure of neurons in a particular context in the brain that when present in the right number and relationship might attain consciousness.[6]

That consciousness emerges from cells is an *assumption* many scientists hold; it has hidden within it another fundamental assumption—that (eventually) science will be able to explain everything. Because if science can explain consciousness, science can explain it all. It is only a matter of time before humans will from scratch recreate consciousness, creativity, and imagination in a computer, and in fact, a lot of resources are being spent right now toward this end.[7] Man equals machine.

The Romantics and many of their ideological brethren take issue. The self is unknowable by the self, and actually, as elaborated by physicist Sir James Jeans, "The material universe is derivative from consciousness, *not consciousness from the material universe* [emphasis added] . . . In general the universe seems to me to be nearer to a great thought than to a great machine."[8] But to many scientists, claiming *anything* is unknowable by science is not consistent with their worldview. When pressed, some of my neuroscientist colleagues, while comfortable with the notion of the mapping of *physical* activities onto neurons, are less comfortable with mapping complex mental capacities such as empathy. And they are even more uncomfortable when I point out that for them to maintain a consistent worldview, it's difficult, if not impossible, to have it both ways.

The Romantics' view that consciousness and all reality arise from the mind is much closer to the Buddhist conception of things. The material world is impermanent, a delusion, a creation of the mind. As Konchok notes and the Dalai Lama elaborates, continuing from his discussion of the flower as something scientists can dissect and study "to the point where we can no longer point to a taste or a smell . . . yet we cannot deny the existence of flowers and their sweet scent": some Buddhist philosophers say the flower exists only as an object of our perception, that each of us sees and smells a different flower. This challenges objective truth, while bringing consciousness to the fore. And, the Dalai Lama adds: "Even the mind is not in and of itself real."[9]

The major Buddhist schools have differing viewpoints on consciousness, with some similarities to and some differences from Western beliefs. The Mahayana School (proponent of the Middle Way) asserts that everything is not established from itself; nothing exists in and of itself, rather it is the result of projections of consciousness.

Another school of thought says that consciousness arises from a physical object. Consciousness and object have a causal relationship. This school holds the view that consciousness of the five senses is produced by the power of their own physical sense organs. For example, eye consciousness or eye cognition arises from the sight sense organ.

Each sense has its own kind of consciousness; there is no central brain. This school of thought says there is a causal relationship between a vase and the cognition that apprehends the vase. This means the vase is the cause, and the conceptual consciousness apprehending the vase is the effect or result. The conceptual consciousness apprehending the vase emerges from the vase. Therefore, according to this interpretation, it would appear that things do not arise from consciousness, but rather vice versa. This becomes very complicated, circular, when considering the things, say neurons and the brain, that science considers give rise to consciousness itself.

So how does a project such as the Emory-Tibet Science Initiative best facilitate collaboration and potential transformation in such discussions? B. Alan Wallace—an American educated in the West, a former monk who studied in a Tibetan Buddhist monastery, and someone we consulted early on in our project—captures the potential. Wallace is writing here about consciousness, but his thoughts extend easily to compassion and empathy and our project as a whole.

> While science characteristically embraces the "disturbingly new," it has a much harder time embracing the "disturbingly old," namely, discoveries that were made long ago (let alone in an alien civilization), prior to the Scientific Revolution. Many Buddhists, on the other hand, rely so heavily on the insights of the Buddha and later contemplatives of the past, that they have a hard time embracing disturbing new discoveries that challenge Buddhist beliefs. Scientific materialists are so confident that the mind is nothing more than a biological phenomenon that they confuse this belief with scientific knowledge. Similarly, many traditional Buddhists are so confident of the validity of their doctrine that they confuse their belief with contemplative knowledge. . . .
>
> The scientific tradition has now joined the Buddhist tradition in its pursuit of understanding the nature, origins, and potentials of consciousness. At this point in history, it may be said that neither embodies a rigorous, unbiased, multifaceted science of consciousness. But as scientists and Buddhists collaborate in the investigation of this phenomenon so central to human existence, perhaps such a science may emerge to the benefit of both traditions and the world at large.[10]

To sow, nurture, and plow this field, as Wallace suggests, requires significant give and take from both the scientific and Buddhist worldviews. What does each worldview bring to a discussion of compassion and empathy?

If the core of Buddhism is the truth of suffering, its source, and how to relieve it, then certainly compassion and empathy are at the heart of that mission. And as far as tools: Buddhists, as Wallace points out, have developed over centuries meditation practices—first-person, inner "science" or techniques that foster this mission. And perhaps such techniques are what has provided Ngawang and the diviner with their extraordinary capacities for these traits.

Science brings a truckload of techniques and approaches, virtually all of them based on third-person observation, to a large extent separate and disconnected from the object of study—in this case the brain. Such techniques should complement the first-person approaches of Buddhists. How can they be used to search for and measure differences in mirror neurons or some other activity in Ngawang, say?

If Buddhism and science are to come together and rethink science to allow less separation and less fear of contradiction, to engage the best of science and Buddhism to understand what makes us care for each other, and to make compassion and empathy integral to both world-views, then scientists, like Buddhists, should realize they probably do not have all the answers and should be open to a bit of enlightenment (with a small "e"). One approach is to open some conceptual space: rather than thinking a prescribed combination of cells and molecules *equals* thoughts, emotions, and beliefs, for example, consider instead cells and molecules as the substrates, *the materials through which* the environment and experience move, necessary physical entry points.

Before exploring mirror neurons and other brain cells, we return to the beginning, where all biology stories start: evolution.

BOTH THE DALAI LAMA and Darwin make it clear that the evolution of emotions is an important and engaging topic. Darwin wrote a whole book on the topic: *The Expression of the Emotions in Man and Animals*,[11] and the Dalai Lama sees the evolution of emotions as a superb conceptual entry point into a biology-Buddhism collaboration.

We know we are *seeing* compassion and empathy when Ngawang wishes for a Thanksgiving shared with Chinese (he has since established such an annual event at Emory), when the diviner grants grace to his torturers, or when the Dalai Lama interrupts his own speech to thousands in Atlanta to leave the stage and hug a hero of his he spots in the front row.

Over dinner the evening before, Richard Moore, this hero, related his unbelievable story in a singsong Irish brogue. At age ten, Richard was left permanently blind by a rubber bullet shot at him from point-blank range when he was on his way home from school. This was during "the Troubles" in Northern Ireland more than four decades ago. Richard's distraught father asked the doctor that day if he could donate his eyes to his son.

Our food remains untouched. Richard tells us he never bore any ill will toward the British soldier who shot him. Richard went on to raise a family and become a successful businessman and then started a foundation that assists children worldwide caught in the crossfire of war. Astoundingly, as an adult, Richard sought out the man who shot him, forgave him—and became his friend. Moore's story, also told in his book *Can I Give Him My Eyes?*,[12] recalls the diviner's story.

In order to study and measure such empathy, to analyze it in evolutionary terms, more definition and concreteness are required.

If empathy is a complex process that involves sharing an emotional state with another organism,[13] compassion can be defined as actions taken based on empathy, a kind of advanced altruism. Resonating with the stories of the diviner and Richard Moore, the Dalai Lama says, "True compassion is not just an emotional response but a firm commitment founded on reason. Therefore, a truly compassionate attitude towards others does not change even if they behave negatively."[14]

The Dalai Lama and evolutionary biologists agree that the origins of these emotions probably arose from the relationship between mothers and their offspring and, in the case of mammals and many other animals at least, the absolute dependence of the child on the mother.

Compassion and empathy are not unique to humans, and as we will see, it is becoming more and more difficult to draw lines between these traits in humans and in other animals. The Dalai Lama calls interdependence—the core from which compassion and empathy arise—a "fundamental law of nature" and notes that even some insects are social beings "who without religion, law, or education survive by mutual cooperation."[15]

So in general, there is consensus between biology and Buddhism that emotions are complex traits which evolve like other complex traits. But the devil's always in the details.

When I first learned that emotional states such as aggression, anger, desire, and fear are necessary, from the evolutionary perspective, for the survival of the species, I found it challenging to accept. In Buddhism, these emotions are considered afflicted and *dis*advantageous, and therefore we need to remove them. But based on scientific thinking, these traits evolved and are advantageous to organisms. For example, perhaps if you are more aggressive you will be more likely to survive in the population and probably get greater access to resources.

But then I saw a resolution to this challenge. Evolution does not classify traits such as these emotions as good or bad; human cultures do this. Actually, each of these emotions can be good or bad, negative or positive, depending on the situation. The negative should always be rejected and the positive adopted.

The societal context of both Buddhism and science, in intriguingly different ways, affects discussions of the evolution of emotions. Konchok resolves this tension between Buddhism and biology by realizing that evolutionary ideas *do not assign value*.

Ironically, as my colleague Frans de Waal often relates to my students, it is taking scientists a bit longer to come to such a resolution, especially where "positive" emotions such as empathy, love, and compassion are concerned. de Waal is a primatologist and studies empathy and its evolution. Even though Darwin thought and wrote about such emotions, as de Waal notes in *The Age of Empathy: Nature's Lessons for a Kinder Society*, what scientists (and most of society) first took and ran with from Darwin was quite different. They instead focused on and embraced his themes of survival of the fittest and humans as aggressive competitors. Darwinism was taken up as a justification and support for capitalism, the predominant economic movement—in fact, dubbed "Social Darwinism"—in the British empire at the time that grew and spread in the late nineteenth and twentieth centuries.[16]

So in comparison to Buddhism it is *exactly the opposite* prejudice that scientists and Western society often bring to the discussion. As de Waal says, even in the 1980s when he first started studying empathy (only because of a striking event that he witnessed among a group of captive primates in the Netherlands) he was getting funny looks at scientific conferences; there was a lot of resistance to such work.

Science shapes and is shaped by culture in powerful ways. The scientific establishment wondered: What is the point of studying positive behaviors and emotions? This challenges the Western narrative (de Waal goes so far as to call it the "macho myth") that humans are inherently independent, aggressive, competitive, war-waging animals. Empathy and compassion are soft, subjective, nonscientific human creations. Western science gravitates toward studying what goes wrong, the negative, the mutations, rather than what prevents bad things from happening in the first place.

But de Waal and others make a strong scientific case—complementing the Dalai Lama's—for the evolution of empathy and other traits that increase the success of the group. Yes, of course, competition and aggression are important, but cooperation is also vital. The most successful individuals are the ones who best read and respond to the emotions of others, whether those others are in the family or outside it. For most of human history, before we took up agriculture and settled down, life was about the importance "of reliance on one another, of connection, of suppressing both internal and external disputes, because the hold on subsistence is so tenuous that food and safety are the top priorities." The idea that humans have always been violent and engaged in tribal battles is inconsistent with the fact that modern humans are revolted by killing each other (even our enemies) and suffer greatly when we do.[17]

As with all good evolutionary arguments, researchers must move beyond logic and just-so stories and, if possible, test their ideas in current species. If empathy is a beneficial trait, it should have evolved like any other such trait. Evidence for conserved elements of this trait should be found in other species and at all levels of the Living Staircase—communities, individual behaviors, brains, cells, and genes. Indeed, among primates, dolphins, and wolves, more evidence exists for cooperation and reconciliation than for war and savagery.

KONCHOK REFLECTS that all sentient beings "have the same emotions." But rigorously studying emotions in other animals (or humans for that matter) has thus far proven quite difficult. While situations often can be more easily manipulated with animals, the animals of course cannot self-report; thus, most scientists are very careful about using language that over-anthropomorphizes animal behaviors. Also, virtually all of the "animal emotion" studies carried out are done with animals in captivity. Nevertheless, some vestiges of empathic-like behavior, at least, do occur in other organisms.

Apparently, the old story about elephants being afraid of mice is not true, but mice do demonstrate at least some "primal empathy" toward *other mice* experiencing fear; and recent work on elephants and nonhuman primates suggests these organisms move well beyond that, acting to calm stressed members of their social group.[18]

Here we get to some measurables: mimicking behaviors, acting to help some of the species but not others, and sharing-behaviors with some but not others. Do researchers see activity in similar brain regions of the same and different species that are engaged in such behaviors?

When one mouse observes another mouse receiving electric shocks, the observer mouse also shows physiological signs of fear, freezing in place. Also, when a day later the observer is placed in the same location where it observed the shocking of another, it again freezes as if it, the observing mouse itself, were fear conditioned. As always, context is very important here; the responses of the mice are very different depending on the nature of their relationship with the animal they are observing. These effects increase with familiarity; if the observer involved is a mate or sibling of the mouse being shocked, the observer demonstrates more intense fear behaviors than it does for a nonrelated mouse.[19]

By genetic and pharmacological manipulation, researchers show that the empathic reactions of the observer mice engage the same *area*, not necessarily the same neurons (it's not clear yet if this is a limitation of the technique) of the brain that is activated when the observer itself experiences pain. So perhaps the observer mouse literally "feels the pain" of the mouse it is watching. Analogous areas of the *human* brain are also active in empathic situations.[20]

Bartal and colleagues show that rats can take things one step further. Rats observe a fellow rat in need and then *act on* that perceived need. Again, the behaviors are context dependent. Rats free cage mates from a restraint, but only if they have spent time previously with the restrained individual; they do this irrespective of genetic relatedness. The rats free cage mates even before accessing chocolate, so clearly a powerful force is at play here.[21]

Asian elephants also practice such "targeted help," assisting injured or dying group members. de Waal and collaborators measured what they call "reassurance" behaviors among a group of elephants. Once an animal demonstrates typical distress behaviors, researchers record the nature and timing of the response of bystander elephants. Within one minute of the distress, bystanders commonly provide unsolicited affiliative contact with the distressed individual (rather than vice versa)—suggesting the behavior is a kind of consolation. Also, immediately following the distress, bystanders produce specific vocalizations and bunch around the distressed individual with all elephants within touching distance, which may well also serve as reassurance.[22]

Nonhuman primates, dogs and wolves, and some birds also appear to reconcile and console. Via physiological measures (such as levels of stress hormone present), empathic-like behaviors actually reduce stress in consoled nonhuman primates. Primates can also perform "flexible targeted helping"; that is, they can figure out what a member of their social group needs—say, a missing tool—locate it, and provide it. Again, these behaviors are stronger for primates that are interacting with relatives.[23]

Although research by its nature has focused on organisms of the same group and has demonstrated consistently that the more related the organisms, the more empathic they are toward each other, DNA evidence suggests that nonhuman primates in the wild also partner with *unrelated* animals of the same species. de Waal relates a moving episode: he observed a bonobo in a zoo attempt to assist a dazed bird she had captured by carrying it up a tree and spreading its wings.[24] YouTube and other informal sources are replete with images of cross-species caregiving (dog mothers and kittens, a mastiff and a young chimp, an antelope and a lioness, polar bears and dogs, etc.).

The hormone oxytocin is important in empathy in many species. Oxytocin first made headlines as "the love hormone" based on experiments done at Emory on two closely related types of small rodents called voles. Prairie voles are monogamous, while montane voles are polygamous loners. Prairie voles have more oxytocin receptors than montanes in a part of their brains known to be involved in motivation and reward, and when those receptors' activity is blocked, monogamous voles' behavior becomes more like their loner cousins.[25]

The same hormone and its receptor play similar roles in humans; different genetic variants of the receptor are associated with levels of empathy and with some cases of autism. The receptors also provide an intriguing link with stress, as the same genetic variants affect brain structures involved in social threat or stress. Oxytocin is linked with compassion and related feelings in humans. Empathy-inducing film clips correlate with increased self-reported feelings of empathy and a 50 percent increase in blood oxytocin levels, and intranasally applied oxytocin gives rise to a whole raft of compassion-associated behaviors and feelings: increases in trust, generosity, and eye contact, and in ability to read others' emotions, respond to crying babies, and have feelings for criminals. Recent work suggests the increase in empathy induced by oxytocin is for those people already considered part of one's in-group, so it may be as much (or more) tribal hormone than love hormone.[26]

BOTH THE LOGIC and the evidence are convincing that empathic behaviors evolved, and there is a strong suggestion that these behaviors are dependent on environment and context—reminiscent of the context dependence of gene expression discussed in chapter 4. That is, it is not just about sharing an emotional state with another and acting on it, but whether this sharing and action occur in the context of and to what degree depend on, among other things, the relationship between the organisms.

If empathy depends on the environmental context, is it possible Ngawang and the diviner, or anyone for that matter, could *intentionally alter their environment* to increase their capacity to feel empathy and compassion for other people? Could they thus also expand this capacity beyond humans to all sentient beings? What are the cellular substrates of the empathic behaviors scientists observe in so many diverse organisms, and could they be studied as a way into addressing this question?

This chapter opens with the suggestion that electricity shooting through specific cells called mirror neurons might be vital for empathic feelings and actions. Recall that mirror neurons are cells that "fire" both when a monkey does something itself or when it observes *another* doing the same thing. The same neurons seem to fire only when a living being performs the action, not when a robot does. Intentional action by oneself or intentional action perceived in others excite the same neurons.[27] Similarly, in the swamp sparrow, individual cells fire with nearly an identical pattern when the bird hears a song sung by another bird as when the bird sings the same song itself.[28]

Do humans have mirror neurons, and does it matter if we do?

I HAVE NOW WRITTEN a number of textbooks for the monks and nuns in our project. The books have been translated and distributed throughout the Tibetan world to both lay Tibetans and monastics—in both India and Tibet. The text, figures, and figure legends of the books are presented in English and Tibetan on facing pages—a literal and metaphorical representation, a mirror, of our collaboration. When we presented the first of these books to the Dalai Lama, when he held it in his hand and saw the pages, he was quite moved. He seemed as surprised as we were that our project had somehow progressed to the point of having translated textbooks; it made everything more real.

When I set out to write one of the neuroscience texts, I started reading about mirror neurons in top neuroscience journals and got very excited; this would be the perfect lens through which to teach monks and nuns neuroscience. For the textbook, I developed a story about a monk walking through the woods and finding a lost child, feeling empathy and compassion, and scooping up the child. I interwove what might be going on in that monk's nervous system—what neurons are and how they work. Brilliant, I thought. Then I sent the first draft to my colleagues who teach neuroscience to the monastics. I got lambasted: "What is this? Can't we get someone who really understands neuroscience to write this?!" I had unknowingly stepped into a minefield.

Why this is a metaphorical minefield reveals a lot about the nature of Western science and how it drives culture and vice versa, so it is worth tiptoeing through this minefield for a moment. I uncovered some underlying, explosive tensions among the many diverse disciplines that study and have their own views about emotions such as empathy. To try to get a handle on what is driving such tensions, I talked with colleagues in these different disciplines.

THE ENLIGHTENED GENE

The fact that mirror neurons had become *the answer to everything* so quickly really irks basic researchers. A similarly tense controversy with similar roles and results—and the political and personal again greatly influencing the science—is currently going on around epigenetics.[29] Some neuroscientists I talked with have a number of problems, both scientific and philosophical, with mirror neurons. First, from the science perspective. These basic researchers study mice and birds and analyze nerve cells directly; they measure electricity from these cells and construct the nature of organism actions and behaviors as a direct result of these measures. This neuron connects to that neuron, and both are excited when this behavior happens; so these two neurons are involved in this behavior.

Such measurement is difficult in humans, because we don't have the techniques yet to easily and noninvasively monitor single cells in a live person's brain. Scientists studying human brains instead often use functional magnetic resonance imaging (or fMRI) to correlate neurons to behaviors, actions, and thoughts. This method provides very striking and colorful images of the brain "lighting up." What are measured are the metabolic activities of thousands of neurons at once, *regions* of the brain.

Not only does fMRI not allow resolution at the individual neuron level, but as my colleague psychologist Scott Lilienfeld points out, these images are not actually the photographs of the brain they appear to be, but are instead images generated by a computer. The construction of these images necessarily involves significant researcher manipulation and interpretation. Thus conclusions based on fMRIs, while not meaningless, involve correlations of correlations—far removed from the electrophysiologist's recordings of individual neurons in model organisms.[30]

Using fMRI, a method with many problems, some scientists or others interpreting such research have gone on to claim that humans *must* have mirror neurons because a given area of the brain lights up when people are being empathic in fMRI machines. Some then make a premature leap, based on such experiments and despite their limitations, that we now understand empathy, that the "problem" of empathy is solved.

ANOTHER REASON some of my neuroscientist colleagues seem upset about mirror neurons is because these cells became an empathy meme.

The media and the public easily and eagerly picked up on the concept of mirror neurons and the possibility that humans might have them; the Western worldview makes them practically irresistible. What could be more simple, attractive, linear, and reductionistically pleasing than a one-neuron/one-emotion model, complete with colorful brain images "showing" empathy and lit-up "empathy neurons"? We *want* something this simple and easy to understand, and we *want* science to tell us this is how it is.

This scenario is also strikingly reminiscent of the one-gene/one-disease meme that blew through the West during the 1980s and 1990s and still creeps into popular discourse. One peer-reviewed paper after another claimed to identify "the gene for" depression or "the gene for" homosexuality. Implicit was the promise that once the gene for this or that was identified, scientists could then replace a "bad" version of this gene with a "good" one, and all would be well. Very few if any of these claims have come to fruition. Humans are just too complex, and as we have seen, biology is only part of the story. Societal forces, though, are powerful and can circle back to affect profoundly which research projects receive funding and which do not.

Other scientists ask, even if a mirror neuron model is proven, so what? The empathy problem is far from solved. The interesting questions would be and are still far from answered—how do such neurons work together to result in empathy? Are these neurons a cause or an effect of empathy? And, many ask, why the need for "neuroscience's molecular and mechanistic 'high-tech' stamp of approval" anyway? *Psychologists* after all have been proposing for decades that common neural codes exist for perception and action.[31]

Indeed, psychologists such as my colleague Philippe Rochat note that a key to understanding empathy is to look at it from the developmental psychology perspective. Humans show different levels of empathy at different biological ages. At birth, humans have an "automatic" empathy; by two months of age, infants reciprocate gestures and share experiences through imitation and games; around nine months of age, infants share with others their relationships to things in the environment and move beyond face-to-face exchanges; by fourteen months, humans have a strong sense of self, recognize themselves in the mirror, and can project themselves onto others; around the age of two, capacities mature for feelings based on an understanding of relation of the self to others; and by age four, the ability emerges to consider the minds of others, to think about what others are thinking and feeling.[32]

IN THE END, the monk textbook was rewritten with an extremely limited discussion of mirror neurons, but with significant science learned and cultural insight gained. As long as we keep all the stated caveats and limitations in mind, it's still very much worth reviewing the work of neuroscientists using fMRIs and other techniques in relation to our quest here: searching for contextual influences on empathy and the brain.

The work of the mind should engage a diversity of fields in the West —neuroscience, philosophy, psychology, sociology, and anthropology, for example—much less the rich cultural traditions of the monastics. It is clear only that things are not clear in the mirror neuron field, that when we dig into it, as with much of science, we learn that the knowledge is intriguing and increasing, but also interpreted differently and shaded by expertise, expectations, and culture.

Could emotions and feelings such as empathy have evolved like any other trait?

On the one hand, I would say yes, because generally I consider that most human traits develop from different causes and conditions within environmental, social, and cultural situations. Similarly, emotions and feelings of empathy are generated from quite similar conditions as other traits.

Depending on the situation, various causes, conditions, social situations, and the environment can generate feelings of empathy. For example, if you were strolling down the street and saw a serious accident, you probably would see badly injured people and other witnesses shouting for help. These situations almost always arouse powerful and empathic feelings in people. Most people would automatically sense, understand, and share those people's emotions, feelings, and reactions of empathy that were generated by that powerful situation.

Some of my fellow Buddhists believe differently—that emotions and feelings such as empathy do not evolve like other traits, because empathy is more a state of mind, which is learned or can be promoted by certain practices.

According to the knowledge that I have gained from studying evolution, together with neuroplasticity, neurogenesis, and epigenetics, I think thoughts and feelings such as empathy and compassion could possibly be reducible to neurons, other cells, and genes. It is clear that compassion and cognition are interconnected with neurons, other cells, and genes. For instance, if a person is unhappy, then you might find some problem related with the emotion of happiness in the brain; if something is not functioning well in the brain, the results might manifest in the person's emotions.

I had a discussion with some of my Buddhist friends about compassion and emotion. During the discussion, just for fun I said, "Emotion is a by-product of the electrochemical activity in the brain."

One of my friends protested, "I don't agree. If that is the case, then one could measure emotions. Can we really weigh and measure emotions?"

His logic was off the mark because brain activities *are* measurable. So I answered, "Yes, you can measure emotions through brain activity."

My friend became angry and rebutted, "Your concept is against our philosophy."

I thought: I am not so sure.

As we have seen in our exploration of resilience and epigenetics and throughout this book, humans and other organisms have evolved to respond dynamically to the environment in the short and the long term. Such responses are apparent along the Living Staircase from genes and cells to brains and behaviors. Change in the brain and its genes and cells—in both structure and activity—correlates to change in *social* environment. As we saw, mice that experience different levels of care early in life have as a result different levels of gene expression that are correlated to changes in behavior later in life under stress. In humans, early stressors appear to affect the size of certain brain structures, such as the amygdala, correlated with stress response.[33]

In addition, social context has a significant impact on empathic behaviors—even among rats and mice. In humans, preliminary social neuroscience studies suggest an impact of social experience at the brain level. Empathy and related behaviors are affected not only by genetic relationships, but also extend to historical and more complex social connections. For example, in one experiment Jews were told that half of the people they observed—all Caucasian with similar physical features—were neo-Nazis. The Jews, while in an MRI machine, watched videos of the people identified as neo-Nazis performing simple physical tasks such as drinking from a cup. When a researcher trained in brain analysis viewed *motion*-related areas of the observers' brains, he was able to differentiate between "likable" and "dislikable" neural activity.[34]

This and much other research suggests that human feelings can affect the way we engage and empathize with others, even at the basic level of perception and physical action. Extending these findings to observing pain, Azevedo and colleagues measured autonomic responses (pupil dilation) and brain activity in areas associated with pain in Caucasians and blacks observing Caucasians or blacks in pain. By both measures, activity was greater for pain observed for own-group versus out-group individuals[35]—reminiscent of the in-group effects of the hormone oxytocin discussed previously.

On the one hand, this can be very depressing—people are biased against those not in their group—but on the other, the science says that empathy is malleable; society and culture, many aspects of which humans can influence, make a difference. People, their biology as individuals and in groups, can change and do all the time, every day. Steven Pinker even argues persuasively that the human species *as a whole* has changed dramatically over time, becoming much less violent over the centuries.[36] Experience makes a difference to who you are, whom you empathize with and how much.

Humans have the *capacity*, alterable by experience, to empathize with others, yes in their own group, but with everyone. A striking result about fear among different races found by Elizabeth Phelps and colleagues comes to mind. In these experiments and consistent with the experiments discussed previously, people maintain a conditioned fear longer for those not of their own race than for those of their same race. However, the one factor that mitigates this effect is the experience of interracial dating. In other words: there's hope. Productively and proactively engaging each other makes a difference; we know this intuitively, and the science supports it.[37]

Evolution and biology shape us to be most comfortable, most empathic with those with whom we are most familiar. This suggests humans can intentionally shape the context and environment to affect empathy and perhaps include more of us in our in-group, as Ngawang and the diviner appear to have done. The Tibetan-Chinese Thanksgiving initiated at Emory led to a more formal partnership, the China-Tibet Initiative, unprecedented and so unlikely that the news of it spread throughout the international Tibetan community.

While we were at Emory, we Tibetan monastics helped start the China-Tibet Initiative. This initiative was designed to create understanding between students of the two countries through dialogue and conversation for change and peace.

The China-Tibet Initiative brings new perspectives to international conflict. We tried to build bridges. I believe this kind of initiative moves us in the direction of, as His Holiness the Dalai Lama has mentioned, making the twenty-first century one of dialogue, peace, and understanding. I also believe Emory's China-Tibet Initiative is an inspiration for other Chinese and Tibetans worldwide to reach out to one another in a similar effort of friendship.

During meetings, we shared personal stories and ideas, and every other weekend we monks led meditation sessions with students involved with the initiative.

It is precisely through these connections and by creating an environment for safe and open dialogue that we can work together to solve many problems—from global warming to treating disease. We can also explore the perceived contradictions between science and religion. It is such collaboration that can solve many misunderstandings and move both cultures forward.

Globalization is forcing interactions across all cultures. Perhaps humanity's best hope is to take advantage of such opportunities to invite each other in, as the Tibetan Buddhists have invited in Western science, gain familiarity with each other, search out commonalities, limit biases, and increase empathy. This is at the heart of the Dalai Lama's idea of secular ethics that he discusses with us in chapter 8 and in his book *Beyond Religion*.[38]

If we think in terms of societal enrichment and empathy improvement, our project could be a model for such cross-cultural fertilization. Between biology and Buddhism, between the third-person and the first-person perspective, we help bring the centuries-old meditative practices to enhance compassion into the West by providing an understanding, for the meditators, of the science that might help investigate their practices in new ways, by helping rethink the way science is done in the lab, and by taking what is learned into the classroom, a potentially rich arena for shifting perspective among the young and aspiring scientists and physicians.

Early signs are encouraging on the empathy research front. Scientists are now beginning to figure out how to measure behavioral and neurological differences in empathy and actions based on that empathy. Several studies examine longtime practitioners of compassion meditation or novices trained in secularized compassion meditation approaches adapted from Buddhist techniques.

Jazaieri and colleagues show that in a randomized, controlled study, novices who practice compassion cultivation training for nine weeks are more compassionate, mindful, and happy; less worried; and less emotionally repressed than controls—this is by self-report.[39]

Novices can learn to nurture feelings of loving-kindness during quiet periods of concentration. In one study, for example, subjects trained in compassion meditation who watch videos of people in distress report feeling better and show stronger activation in areas of the brain implicated in love and affiliation than control subjects.[40] In another, people who undergo compassion training demonstrate more altruism and have altered activation in brain regions associated with social cognition and emotional regulation.[41]

Similarly, at Emory, Mascaro and her collaborators showed that after undergoing compassion-based meditation training (as compared to controls), subjects both score better on psychological empathy tests and have increased activation of areas of the brain associated with empathy response. The increases in brain activity are compared to individuals' baseline activities prior to the start of the study interventions.[42] Richard Davidson and colleagues have also found alterations in response to pain and emotional sounds of distress in the brains of novices versus long-term meditators.[43]

Buddhists and some biologists accept that meditation is very useful and can improve empathy and compassion. Most Buddhist meditators see and understand meditation's benefits, but most do not realize how it works biologically. Myself, I have experienced a little bit of meditation's benefit through my own practices, and furthermore, I have read many articles and heard from my guru about the value of meditation.

Meditation has improved my concentration and decreased my rumination. Scientists find that physical changes—such as increased gray matter in the hippocampus, which is related to positive emotion—take place in the brains of those who meditate regularly. Likewise, meditation increases gray matter in parts of the brain associated with positive emotions such as compassion and empathy.

Both compassion and empathy suggest the capacity to share another person's feelings, needs, and emotional state. If one is not well trained, one cannot perform a genuinely selfless act. Genuine selfless acts can be shared or performed only by people who have achieved a certain level of practice.

We know that every experience and new habit impacts the structure of the brain. Therefore, empathy is an experience that will impact the structure of the brain, and that experience can be improved and made more profound through meditation practice. The more you train, the more you have experienced. The more you have experienced, the deeper your insight and the greater the physical changes in your brain.

Buddhism says meditation can deepen insight and sharpen the mind, bring more awareness, more calm, and less aggressiveness. Also, meditation is very helpful for other health issues such as blood circulation and heartbeat.

Meditation has improved my concentration and helped me in not making things such a big deal.

In the next chapter, we explore similar studies and techniques in relation to their effects on disease and stress response; that is, compassion interwoven with physical and mental health.

TO EFFECT real change, we must start with the children. Here the early news is good. In a meta-analysis of some 270,000 children in 213 kindergarten through twelfth-grade programs promoting emotional regulation and prosocial behaviors, students not only made gains in social skills but—reminiscent of other enriched educational approaches discussed in chapter 3—also scored 11 percent better on standardized achievement tests as compared to controls.[44]

Barbara McClintock, Nobel laureate—and after Mendel, perhaps the most famous plant biologist of all time—embodied contemplative science, an intertwining of research, learning, and empathy. In her aptly titled biography, *A Feeling for the Organism*, McClintock often speaks of the corn in which she did her groundbreaking work on "jumping genes" (similar to the LINE-1 elements later discovered in animals, and discussed in chapter 4) almost as a friend. As she stated publicly only later, McClintock actively worked between biology and Buddhism long before it was a part of mainstream discussion in the West.[45]

McClintock says she always had an "exceedingly strong feeling" for the oneness of things. "Basically everything is one. There is no way in which you draw a line between things. What we do is to make these subdivisions, but they're not real. Our educational system is full of subdivisions that are artificial, that shouldn't be there."

McClintock said she knew and discovered things about corn and genes "in an internal way." And that after she knew, "What you had to do was put it in their [other scientists'] frame . . . you work with so-called scientific methods to put it into their frame after you know. Well, [the question is] *how* you know it. I had the idea that the Tibetans understood this *how* you know."[46]

How do we connect the first- and third-person perspectives, most effectively break down the subdivisions of which McClintock speaks, and bring empathy into the science classroom and lab? How do we best create a science in which researchers like McClintock may not have to translate their knowledge into science-speak?

Back in the United States, students in the "modern" science curriculum, which focuses on information, some critical thinking, and on preparation for medical school or entering other areas of health care, have forgotten all about plants—their essential biologic relevance, the parts of flowers, much less their beauty and scent.

I hope that, unlike the monks and nuns, my students are not surprised that plants are living, responsive organisms that share a common ancestor with themselves. But I am not so sure, so I always bring my biology classes outside to sit in the grass, wonder at the trees, explore the cells of plants, pull apart leaves and grass and flowers, and wrestle with problem sets about plants and their cells and molecules. I do, at first, sense a kind of disbelief and discomfort among the students. They are more comfortable being separate from the outside.

My students are lucky to have on campus a beautiful park complete with lake, trees, fishing, and walking paths, basking turtles, the occasional deer, fox, or beaver. Here is where my classes go. Once we get down by the water, the students relax and get to work. They quickly discover—as did the monks dissecting their flowers in Dharamsala—that plants face and solve many of life's problems with similar strategies and even similar molecules as they do. Years later they remember these outdoor class sessions—molecules and flowers and the outdoors linked.

Arthur Zajonc and others (recalling McClintock) call the integration of empathy and compassion with science and education "contemplative inquiry." Zajonc ascribes to such inquiry eight characteristics in a speech he gave at Emory: (1) respect ("stand guard over and protect the nature of what is being studied"), (2) gentle empiricism (be gentle so as not to distort the object of study), (3) intimacy (rather than disconnect from the object of study, get close but exercise care and restraint), (4) vulnerability (be open to that being studied and ready to admit lack of knowledge, contradiction, and ambiguity), (5) participation ("in meditation we join with the other," analysis is experienced in and through what we are studying), (6) transformation (we internalize what is experienced through the other and this changes us), (7) formation (we change the way we live and see the world), and (8) insight (called "direct perception" in Buddhism, this is seeing without reasoning getting in the way—perhaps what McClintock was referring to when she knew things in that "internal way").[47]

And thus the challenge: How to capture for my Emory students the glee of the monastics the first time they went into the Himalayan foothills and experienced the outdoors, as Konchok says, as *both Buddhists and scientists*. And how to ensure that the monks and nuns, as they learn more science, do not lose that glee.

CHAPTER SEVEN

Meditation and the "New" Diseases

Let's begin with three full breaths . . .

GESHE LOBSANG TENZIN NEGI,

COGNITIVE-BASED COMPASSION TRAINING

[quoted throughout the chapter]

vibrant cheer rises from the twelve hundred graduates, rises past the thick, majestic white oaks, and heads skyward from the quad on this surprisingly cool Atlanta May morning, the day after Mother's Day 2013.

"We honor six Tibetan Buddhist monks for their successful completion of courses in the college over the last three years," says the dean of Emory College, decked out in the grandly ridiculous formal robes of academe. "These Tenzin Gyatso Scholars are here as leaders in the Emory-Tibet Science Initiative. . . . And they now return to their monasteries in India to become science educators themselves. The monks have become part of the fabric of the Emory intellectual and social community. We honor them today with a special certificate."

For the monks, the university, and our project, it is a landmark day. Konchok and his colleagues, all of whom, in addition to being the first alumni of the five-year science course in Dharamsala, have now also just completed three years of undergraduate classes at Emory in the natural and social sciences.

As the college graduation ceremony begins and Konchok and his five monk brethren rise from their front-row seats one by one as I call their names, to receive their certificates, it is that spontaneous cheer from the students that hits me, makes it clear the dean's comments about the monks' impact on the community are more than mere words. For not the first time in my years working with this project, and as corny as

189

it sounds, I am flooded with that weird warmth that comes with being part of something that might matter.

Walking onto the stage, receiving the certificate in front of thousands of graduates and guests, deans, members of the faculty, and the commencement speaker—former US poet laureate and Pulitzer Prize–winner Rita Dove—was special, exciting, and a great honor. The graduation was one of the most memorable days of my life.

Up there onstage, I remembered the first time I had stepped into a classroom at Emory. I was in awe of the atmosphere. The room was buzzing with excitement and expectation, and I felt both nervous and curious at the novelty of the situation. I walked into a big classroom for the first time and felt intimidated. I was not at all ready to sit down in the front row or in the middle, so I just walked gently and bowed down without looking left or right and sat in the last row.

Nearly every inch of the Emory University quad has been coated with white chairs—fourteen thousand in all, crammed in neat rows, meticulously aligned, tied together by staff volunteers and guarded closely against the elements (or mischief-minded undergrads) for the past forty-eight hours. Now the chairs are filled with the graduates, their families, the faculty, and the dignitaries who pepper such events.

As I look out over the quad from the stage, I see Geshe Lobsang Negi sitting in the white chair on the far left of the front row, next to Konchok and the other monks. Few would guess, I reflect, that Lobsang, this man wearing a Western-style suit and tie, as always quiet and understated, is the key bridge, the crucial link that allowed our project to happen, sparked its beginning and evolution, drives its twists and turns, embodies its hopes and goals. Few would guess that Lobsang grew up in the farming villages of far northern India—more than 7,500 miles away and squeezed between Tibet, China, and Pakistan—dreaming of becoming a monk. A dream that at first seemed impossible because after 1959 the monasteries in Tibet were inaccessible, and new monasteries were yet to be established in the border regions where Lobsang lived.

Years before, in the early 1970s, the Dalai Lama sent out emissaries to the far reaches of the Tibetan communities in India to recruit a new generation of monks to lead his exiled community. It had not taken one of the Dalai Lama's emissaries long to recall and then recruit the young boy who often snuck into his teachings to listen from the back of the room. Lobsang, who was fourteen at the time, went on to become part of a group of young monks—now many of whom are *geshes*—who are leaders in our project and the Tibetan community in exile. Another is Geshe Lhakdor, director of the Library of Tibetan Works and Archives and lead partner in India of our project; another directs the Institute of Buddhist Dialectics in Dharamsala; two others, Tsondue Samphel and Geshe Dadul, are the lead translators in our project.

I think, on that graduation day, of the somewhat karmic connections between Lobsang and me. As we cotaught American undergraduates in a course on mind-body medicine in Dharamsala, we discovered by chance that in 1988, twenty-five years before, he and I had been in the same room—a grand theater in Seattle. Lobsang was one of a number of young monks on a tour (partly funded, I recall, by Mickey Hart, drummer for the Grateful Dead) bringing traditional Tibetan music and culture to the Western world. I was a graduate student at the University of Washington struggling through a PhD in biochemistry and largely oblivious to things Tibetan or Buddhist, or actually to most everything outside of graduate school. The monks' performance was my first date with the woman who would become my wife.

LOBSANG IS MORE than an accomplished scholar and a bridge between two disparate traditions. He is one of the first Eastern monastics to be both a monk and a collaborator *with* (not a subject *of*) Western research scientists. He thus exemplifies a new and hoped-for step in the continuing evolution of the Emory-Tibet Science Initiative and in the developmental arc of such collaborations in general.

Lobsang does research and publishes papers, together with colleagues in psychology and psychiatry, in the peer-reviewed scientific literature. Lobsang has developed a meditation-based therapy, Cognitive-Based Compassion Training (CBCT), and studies its effects on compassion, stress reduction, and the immune system from the molecular to the behavioral level.

In what was initially a response to a request from an Emory undergraduate distressed by the increasing occurrence of depression and mental illness on campus, Lobsang extracted and adapted CBCT from ancient Tibetan Buddhist *lojong* meditative practices, removing religious references or elements.[1] Such secularization is a theme we will see again in chapter 8 in our discussions with the Dalai Lama.

Lojong practices use an analytical approach to cognitively explore unexamined thoughts and emotions toward the goal of developing compassion for all fellow sentient beings. *Lojong* practice begins with challenging the common notion that people divide those with whom they interact into friends, enemies, and strangers. This engagement is followed by learning methods to develop spontaneous love and empathy for a wider and wider circle, beginning with oneself, moving to friends and family, and then to those previously disliked.[2]

DOES Cognitive-Based Compassion Training work?

All of us live in the thick of biological, cultural, and historical contexts. We have seen the powerful effect of societal narratives throughout this book and in at least three examples quite explicitly: (1) in Westerners approaching biological development, from fertilization to death, more as a progressing linear process than as a circular interchange of life and death; (2) within the trope that humans are naturally competitive, aggressive, and war-mongering creatures that Frans de Waal came up against in his research suggesting humans are just as naturally also cooperative, empathic, and compassionate; and (3) in the discussion of mirror neurons and their strong appeal to the linear, reductionist Western scientific view of the world.

In her book *The Cure Within: A History of Mind-Body Medicine*, Anne Harrington strikingly illustrates how the contexts of biology, culture, and history intertwine with and affect each other. My students read this book to start my course "Science and the Nature of Evidence: Why We Believe What We Believe."

Harrington uses the history of mind-body medicine in the West to elucidate how social context and belief alter people, their definitions, beliefs, *and actual physical experience* through time. As Harrington puts it: "Bodies participate in history, even if historians [and here we might add 'scientists and physicians'] do not always know how to deal well with that fact."[3]

Harrington explores American mind-body history through six underlying narratives. Lobsang, Konchok, and I, as well as the Emory-Tibet Science Initiative itself, are part of what Harrington calls the "Eastward journeys" narrative of mind-body medicine. The narrative goes like this: The stress of modern life has damaged us physically and mentally, and modern medicine deals poorly with the results. But not all is lost; there are those in "the East" who can help, because, the story goes, they are the "anti-Westerns." Westerners are stressed and overworked, while Eastern peoples are calm, wise, and attuned to ancient knowledge; Western medicine treats people like machines, while those from the East take a holistic approach in which mind and body are connected—to each other and to the greater universe.[4]

Now for Westerners to gain (or regain) the ancient knowledge and cure our diseases, Harrington continues, we must journey to the East or vice versa—exactly as has occurred in our project. For centuries, Westerners tended to speak and write, in patronizing fashion, of those in the East as the Other, the exotic, the inverse of Westerners. This Orientalism (as it was famously named by Edward Said) evolved in the nineteenth century into a romantic version in which the East was rich in ancient spiritual and moral ideas and ideals that the West had lost and should strive to regain, and then in the twentieth century, into a slightly different version with a focus on both the spiritual and the medical.[5]

It is the middle of the night at Gandhi International Airport in New Delhi. I am on my way from my monastery in the Himalayas to teach and translate science at Gaden monastery in south India. As is typical in India, one of my flights has been delayed four hours, the other six.

An older man approaches me. He tells me he has a serious problem with his knee; he had lost his wife very early and not had children. He asks me about my practices: How do you meditate? Do meditation and yogic practices really calm your mind? Do these practices really benefit your physical body?

My friends back at Emory and people from the West who visit my monastery often ask me similar questions. What tools do you use to cope with certain problems? Which are the best practices for not going back to your early childhood problems?

In my mind linger two questions: Why does everyone always ask me these same things? Are they testing me?

Maybe people think we monks possess some supernatural power or energy that they want to gain from us? From my knowledge, this is not true, but it seems like they do expect this from us.

What can I do? I suggest very simple things to do and practice: do not stay alone when you feel stressed; be with your friends, go for walks. I do not have any special power or energy to solve their problems. I have no special medicine to ensure their speedy recovery. All I can do is share my own experience that I have gained from my normal daily practices. I can show them how to meditate and perform other practices that might help them to get rid of their problems and to live freely. I never forget to share some funny jokes, too.

It is quite gratifying to meet these people, and at the same time, somewhat intimidating to be asked to help those who are going through stressful lives or mental illness.

Harrington traces the narrative arc of Eastward journeys from the Maharishi Mahesh Yogi in the 1960s to the Dalai Lama in the 2000s, when *The Cure Within* was published and just as our project was beginning. Each point along the arc, continues the narrative, features—in addition to a journey or two and "wise men" from the East—Western scientists uncovering "ancient secrets" or the biological mechanisms that enable them. But each point also is rich in cultural and historical significance. The Maharishi and his transcendental meditation (another secularized technique, like CBCT) were, for example, brought to prominence thanks to a famous journey to India by the Beatles.

Bill Moyers helped "invent an ancient tradition" with his 1993 hit public television series *Healing and the Mind*. In one segment of Moyers's show on the mind-body medicine of China, he journeys to Beijing to learn about the connection between human health and the invisible life force known as *chi*. Harrington notes that no images of modern China—huge skyscrapers, etc.—appear on-screen and that Americans at that moment were well primed to accept the idea of a mystical force such as *chi* because of the many references to that concept in the Bruce Lee and *Star Wars* movies hugely popular in American culture at the time.[6]

Harrington points out the rich irony of Americans turning to China as part of our Eastward journeys story. In the early 1900s as China was leaving its dynastic days behind and attempting to modernize, the nation was working hard to *get rid of* traditional medicine. Mao (whose role in our specific story we shall hear more of later) and his communist revolution in 1949 initially also bad-mouthed the old ways, but a few years later—for a variety of political, economic, and cultural reasons—he reversed himself and claimed traditional Chinese medicine was important and unique. He then put his minions to work to create a mix of old and new practices that would allow and hold up local practices, and yet still jive with science and communism.[7]

Thus, the "ancient traditions" around *chi* in the PBS show that captivated the United States were really an amalgam of old and new knowledge, most of which was originally not even linked to medicine per se and had been created only in the mid-1950s. As Harrington relates:

> All this matters, because it highlights a fundamental ethical instability at the heart of the "Eastward journeys" narrative genre. In telling "Eastward journeys" stories, we variously look to India, to China, and . . . to Tibet to function as our Other. The East, though, is not really our Other, and never was. Therefore, in each story we orchestrate, actors must be found to play the role of ancient wise man or ancient healer. Some of the people we recruit to play that role may indeed be wise and may indeed have things to share, but all are also real people, who come from countries with histories at least as complex as our own. In this sense, ironically enough, "Eastward journeys" stories rarely, if ever, take us into another world; they just take us deeper into ourselves.[8]

In the Eastward journeys narrative, Lobsang Negi follows—with an important plot twist—in the footsteps of "the scientist" who translates the "ancient wisdom" into "Western." Herbert Benson was among the first such translator-scientists. Benson is a Harvard cardiologist who was approached by followers of the Maharishi (who himself actually grew up in a middle-class Indian family and received a secular education) and transcendental meditation to study the physiological effects of that practice.

Benson was struck by what he learned, but as a young professor, he was too nervous to publicly reveal who his collaborators were. He wound up stripping almost any mention even of "meditation" from the biological phenomenon he discovered. Benson called it the "relaxation response"—a direct biological inverse of the fight-or-flight response —and published a book in 1975 by that name that has sold millions of copies and has been translated into more than a dozen languages.[9]

A MEETING in 1979 between Benson and the Dalai Lama shifted the nature of the Eastward journeys story and probably opened the way into the space where our project eventually took root. China again rears its historical head.

Benson asked for the Dalai Lama's help and imprimatur to convince Tibetan monks who practice a type of meditation called *Tum-mo* to allow themselves to be studied. This meditation practice is said to allow practitioners to regulate their own body temperature, so they can maintain warmth in freezing temperatures while outdoors wearing little clothing. The Dalai Lama started to say "no" to Benson's request (there was no precedent, and such an undertaking was previously unimaginable), but then suddenly changed his mind, switched from speaking Tibetan to speaking English and said, "Still, our friends to the East [the Chinese] might be impressed with a Western explanation of what we are doing."[10] Here again: religion, mixed with science, mixed with politics.

Another American scientist, Jon Kabat-Zinn, also secularized meditative practices in developing a protocol known as Mindfulness-Based Stress Reduction and has published, beginning in the 1980s, in both popular and peer-reviewed literature on work he has done studying its effects.

In the last several years, Richard Davidson at the University of Wisconsin, working with the Dalai Lama and other Tibetan practitioners, has performed numerous studies of neurobiological effects in long-term meditators, some of which were discussed in the previous chapter. Some of Davidson's research has been done in collaboration with Kabat-Zinn, as well as with Matthieu Ricard, another kind of Eastward journeys translator. Ricard is French and was training as a Western scientist when he left the lab to become a Tibetan Buddhist monk decades ago. Ricard has written and worked extensively in and at the intersection of both areas. B. Alan Wallace, mentioned in the previous chapter, is another Westerner who studied as a monk and is trained in Western science. Wallace now has his own institute in the United States and has published with Lobsang.[11]

And thus Lobsang and his CBCT (together with our project as a whole) continue and extend Harrington's Eastward journeys narrative, taking on the mantle previously borne exclusively by Western scientists. Lobsang and his contemporary Thupten Jinpa, a renowned scholar and translator for the Dalai Lama, are perhaps the only examples of natives of the East who serve within this narrative also as translators *and* researchers.

BACK TO our questions: Does CBCT work? It is evident then that such a question should always be considered in a scientific *and* cultural context. Does it work based on what measurements? Measurements taken by whom? Where? Who is funding the research?

Consider that from 1966 to 1995, nearly all of the studies on acupuncture published from the East showed a therapeutic effect, while only just over half of American studies showed such positive effects.[12] A recent review article in the peer-reviewed literature claims quite confidently that if acupuncture has any positive effect on anything at all, it is so negligible as to be meaningless.[13]

Such variation in results, which is certainly not limited to questions of mind-body or cross-cultural research—are potentially attributable to many factors: the tendency for scientists to unconsciously favor their own hypotheses, the funding source of the research, the tendency for journals to publish only "positive" results that are consistent with the "way of thinking"—the paradigm—of the time, poor experimental design, regression to the mean (only after the sample size becomes large enough, sometimes, can real effects be seen), and other, perhaps as yet unknown, phenomena.

Keeping all these inevitable biases and the narrative(s) in which we are immersed in mind, here the goal is to explore Lobsang's CBCT as an example of a meditation method, an "Eastern" approach, that might help relieve suffering. This exercise aims to work at three levels:

1 To explore from a personal angle an example of how a secularized Eastern approach might work from a biological perspective, and how this can be determined

2 To show how mind-body medicine integrates much of the basic biology explored in the previous chapters

3 To demonstrate the value and challenges of integrating cultures and engaging them in many aspects—learning, teaching, and research—of the phenomena being studied

As Barbara McClintock notes in the previous chapter, Western science divides things up, like the monks' flowers, to study them better. The physical is separated from the mental. The human body is divided into physiological systems—the nervous, endocrine, immune, muscular, skeletal, circulatory, respiratory, and digestive systems. Western medicine is divided into specialized subdisciplines that often reflect these divisions. This allows for patients to access in-depth expertise on specific health issues, but can result in missed therapeutic connections, not seeing the flower.

CBCT and other approaches like it might work and have their effects at the boundaries of and across these artificial divisions. But how?

I meditate—commonly on my own, but sometimes in a group. I used to meditate with my classmates and friends. Meditation became my routine, and I meditate almost every day even if it's only for a short time.

People meditate for different reasons, and the purpose of meditation depends on the meditator. I meditate mainly to get rid of distraction, to keep things in control, to think clearly. When I am mentally foggy, meditation gives me tranquility; I learn how to concentrate long enough to get positive feelings from experiences. Many people like to practice meditation and they want to do it consistently, but I am not sure meditation is for everyone.

In my own experiences, meditation truly reduces stress and other mental discomfort. For instance, it was on February 19, 2009, when I heard my mother died. Soon after I heard the news, I felt very sad. I was tightened by anxiety, was uncomfortable, and lost my appetite. I was totally wiped out, as though from a wave, by such sudden, unexpected news.

Then, to avoid the distress, I started meditation, and I tried not to ruminate on that particular incident. Focusing on my meditation and not ruminating are the best ways to cope with distress. Meditation is a great exercise that can transform mind and body; it is a good method for controlling negative emotions.

The basic idea is as follows. Meditation, and other practices that evolved independently in many different religious traditions, were maintained because they help people think, feel, or act better. From a biologist's perspective, meditation and its kin may well affect the underappreciated but vital conversation that goes on between the brain and the immune system from before birth. This conversation is manifested via inflammation—heat, swelling, pain, and redness.

Perhaps, for example, meditation and its deep and focused breathing stimulates the nervous system. The nervous system then speaks to the immune and stress-response systems in ways that decrease inflammation and improve stress response—a shift in the brain and body.[14] This would help *prophylactically*, in other words, people would have a better response to future stressors, including infection, and *therapeutically*, that is, after people are already sick, their condition would improve. Less suffering.

THIS IS WHERE Charles Raison, a friend of Lobsang's and mine, enters the story. Chuck is an energetic physician-scientist who collaborates with Lobsang in studying CBCT. He is a natural teacher who holds audiences spellbound with his humor, humility, striking ideas, and experimental findings. Chuck was in on early discussions of the Emory-Tibet Science Initiative and taught in it the first year the project was in Dharamsala. That summer was the only time I saw him not pop up and answer his cell phone every fifteen minutes, because as a psychiatrist, he always seemed to be on call for patients, in addition to doing experiments.

Chuck, despite his cell phone, often was a guest in my course on "Science and the Nature of Evidence." In class, Chuck framed the mind-body discussion in Western medicine in terms of the "new" diseases that are now increasingly prevalent in the West—as opposed to the "old," infectious diseases of the past. These "new" diseases are those that Westerners now primarily suffer and often die from. They include diseases traditionally thought of as "physical"—type 2 diabetes and obesity, cancer, sepsis, allergies, atherosclerosis, multiple sclerosis, rheumatoid arthritis, and other autoimmune diseases—but also those considered "mental," such as schizophrenia, autism, depression, and Alzheimer's disease.

Why are these diseases so much more prevalent now in the West than they were just a generation ago? And how do these diseases inform the mind-body discussion? Can meditation and similar practices secularized from religious traditions help relieve the suffering caused by these diseases and also prevent future suffering, as the Dalai Lama imagines? Does research inspired by collaborations from our project and others facilitate understanding of biologic mechanisms underlying such current and future suffering?

The truth is we do not know for sure the answers to any of these questions. The truth is that, similar to studies on acupuncture, one can even find studies which conclude that meditation, for example, rarely works to a significant level. And other studies that disagree with *that* conclusion. This research is in that frustrating, but most exciting phase of complex scientific research where protocols and techniques are being refined and made consistent, where what the right questions are and how to ask them are still being resolved. The whole enterprise is made that much more frustrating and exciting because of the many different disciplines, cultures, worldviews, and perspectives involved.

Here we provide, therefore, no set-in-stone answers, but instead *potential* answers, arising from the work of many, but focusing primarily on those related to the Emory-Tibet Science Initiative. This does not mean these ideas are more or less right than others, but only that they provide a consistent and logical framework for thought.

Let's begin by adjusting the posture—the spine vertical, vertebrae
    stacked loosely, one atop the other . . . hands resting on the thighs or
    lap, shoulders broad, level, and relaxed . . .
Relax the muscles of the face and neck . . .
Check for places of tension and allow these places to soften and feel
    at ease . . .
The posture allows for stillness, relaxation, alertness.

Here's an odd hypothesis: the apparent resilience of Konchok and many Tibetans in our project might be due in part, not just to their meditation practices, but also to their exposure to diverse bacteria, worms, and other microorganisms in the rural, herding, and farming environments in which many of them grew up.

Life is about survival. Humans have evolved to respond to stressors; the better the response, the more likely survival, the more likely a new generation is produced. Major drivers of this evolution are bacteria and other microorganisms. Actually, this is a coevolution.

Recall our question in chapter 1, the lens through which the monastics learned cell biology and genetics: are bacteria sentient beings? Whether sentient or not, bacteria certainly are—for the most part— friends and collaborators in life with humans, a deep and vital part of people, perhaps even integral to our own sentience. Humans "use" these "good" bacteria for innumerable important functions. Bacterial cells far outnumber human cells in humans; bacteria affect human behaviors, help digest food, and influence neural and immune-system development. In turn, these bacteria "use" humans as a safe place to reproduce, live, gather nutrition, and really become us.

To summarize, these bacteria partner with human cells and are collectively referred to as the microbiome. The problem is we did not really know about these good bacteria until the twenty-first century. And because of the heroic work of the microbe hunters, we spent the twentieth century killing them (unfortunately), as well as killing the bad bacteria (fortunately) that kill us. To make matters worse, at the same time, the world was becoming more and more urbanized; thus, in addition to unintentionally killing many good and evolutionarily important bacteria, humans were now exposed to many *fewer kinds* of bacteria and other microorganisms in the womb and during childhood than we had been evolving with throughout previous human history.

What does all this have to do with resilience? And with the possible beneficial effects of meditation?

For initial relaxation of the body and the mind, let's begin with three full breaths . . .
Inhale slowly and deeply through the nostrils, letting the abdomen expand . . . exhaling, releasing tensions and distressing thoughts and feelings, settling into the present moment.

If the immune system is not exposed early in development to what Chuck and others call Old Friends—these microorganisms that humans have shared space with across evolutionary time through soil, dirty water, and domestic and wild animals—the system does not learn to tolerate these microorganisms. When, because of urbanization and antibiotics, humans no longer experience these Old Friends, regulation of the immune system and nervous system development are altered. In combination with particular environmental and genetic backgrounds, this manifests as inflammation, and the result can be increased occurrence of Chuck's "new" diseases.[15] As Chuck himself puts it:

> Because many microorganisms and parasites were always present across human evolution and had to be tolerated, they became "teachers of tolerance." . . . over time the human immune system outsourced some of its tolerance to the Old Friends, an arrangement that worked quite well until the Old Friends were banished from our lives over the last century.
>
> Lacking the "brake" on the immune system provided by the Old Friends, individuals with any genetic risk for heightened inflammation were suddenly left unprotected and vulnerable to whichever flavor of inflammatory or autoimmune condition their genetic inheritance made them most likely to contract.[16]

Think of the interactions among the immune system, the nervous system (plus the hormone system it interacts with), and the microbiome as a three-way conversation.

This triangle monitors the stress response from before birth, through development, and in later life. The three partners "grow up together," learn from, and respond to each other constantly to maintain homeostasis, or return to it, after stress. Ideally, when no stressors are present, the system in the balanced state is fully off; then, upon stress, the response is quick and efficient, and when the stressor is removed, the response is again fully off. "Fully off" in this case means no molecular, hormonal, or behavioral signs of stress are detectable.

Two striking examples—one in mice and one in monkeys—make clear the integral and vital nature of the "triangle of stress" conversation.

In one model system, a viruslike infection in pregnant mice results in pups with many of the same symptoms as some people who experience autism spectrum disorder, including similar cellular and molecular changes and impairment in social interaction and communication.

Already, this suggests a connection between early development and the immune systems of both mother and offspring.

People with autism—and other neurological diseases such as cerebral palsy and major depression—are more likely than those without to have gut disorders such as inflammatory bowel disease. Are the gut disorders in those suffering from these diseases related to their microbiomes? Evidence for this already exists in humans. Is there also a change in the microbiomes of these mice?

Indeed, there is. The newborn mice have gut disorders analogous to those of many humans born with autism; the mice pups have leaky intestines (digestive materials literally leak into the bloodstream) and alterations in the amount and diversity of microbes in their guts—as compared to controls whose mothers were not infected when pregnant.

When one particular species of good bacteria is added back to these sick mice (recalling Marten Scheffer's work, discussed in chapter 5 on tipping-point species in our microbiomes), the changes are dramatic: the gut leakiness ceases, a key molecule called interleukin-6 (or IL-6, known to increase inflammation) is now restored back down to normal levels, and the diversity of bacterial species colonizing the gut increases. These physiological changes are accompanied by changes in the impaired behaviors of the mice. Many such behaviors—including anxiety and communication deficits—return to normal.[17]

A direct cause-and-effect relationship between the nature of the microbiome and the biology and behaviors is apparent. The separation between mental and physical grows fuzzy. The connections here are striking, their exact mechanisms less clear.

Such research also strongly suggests that different species of gut bacteria are, in addition to affecting their hosts, also affecting *each other's* presence and survival, again reminiscent of Scheffer's work.[18] Maybe the good bacteria can secrete molecules that inhibit the growth of or kill the bad bacteria. This has recently been shown to be the case and suggests a novel strategy for identifying new and badly needed antibiotics to use to fight pathogenic bacteria in humans.[19]

Experiments in monkeys also provide strong support for the stress conversation triangle. When young rhesus monkeys are removed from their mothers, the levels of one of their microbiome species, *Lactobacillus*, decrease significantly three days after separation. Again, the stressful event and microbiome shift in composition are accompanied by increased stress behaviors, *as well as increased susceptibility to infection.* The more stress behaviors, the less *Lactobacillus.*[20]

After a week or so of separation from their mothers, during which time the young monkeys establish new social groups with other monkeys, their *Lactobacillus* and stress behaviors return to normal, preseparation levels.[21] Similarly, college students during final exams —presumably times of greater-than-normal stress—have analogous decreases in gut *Lactobacillus* levels (although possible concurrent effects of diet change and other variables are more difficult to rule out in people).[22]

Related studies further strengthen the case for the triangle conversation: mammalian stress signals enhance the growth of particular gut microbes; artificial killing of *neurons* in mice can increase the *growth of the gut bacterium called E. coli* ten thousand times; the stress of sleep deprivation in rats causes overgrowth and redistribution of some microbiome species; cosmonauts in space have altered microbiomes when under stress (as opposed to when not); mice exposed to stressors like social defeat or restraint have an increase in the inflammation-causing molecule IL-6 and a decrease in the amount and diversity of bacteria in their guts. Stressors thus increase susceptibility to infection. Adding back "good" bacteria reverses virtually all of these effects.[23]

Now take a couple of minutes to place yourself in a moment when you felt nurtured. Being immersed in this environment . . . attuning to the feelings of warmth and comfort and the associated feelings of safety and security . . . from being embraced by these circumstances . . . reflect on the value of embodying such qualities as acceptance, kindness, and compassion, so that you may contribute to an environment of nurturance for yourself and others.

Thus, all three participants in the stress conversation triangle—the microbiome, the immune system, and the nervous system—are plastic and dynamic, especially in early development, but also throughout life. In fact, we have already explored the dynamic nature of development and immunity in response to the environment in our discussion of life, death, and development in chapter 2.

Recall the "excess" neurons that are produced during nervous system development and are normally pruned back or killed, depending on which ones connect and where—and how the nature of this dynamic change is often based on the experience, say pairing smell with electrical shock, of the organisms. Many *inflammation-causing* molecules are required for specific types of *learning*, for example, interleukin-1 expression increases in the hypothalamus twenty-four hours after spatial learning, but not before. Depending on context—how much of it is given and where—interleukin-1 can induce memory enhancement or impairment in mice. Interleukin-6, another proinflammatory molecule mentioned previously, is also associated with memory and learning.[24]

Recall also how learning in mammals, even as adults, increases the number and quality of neurons in the hippocampus, and how memory is dependent on both more cells and the death of other cells. The immune system can also affect such neurogenesis. When mice live in an enriched environment, microglia, some of the most important immune cells in the brain, increase in number just as neurons do. The actual structure and number of spines (indicators of memory storage and neural communication) on dendritic memory cells are affected by the inflammatory molecule interleukin-1. In other parts of the immune system, inflammatory molecules suppress neurogenesis.[25] New research on schizophrenia suggests that a molecule previously known to be involved only in immune response also plays a role in limiting neural connections in the brain.[26]

Finally, recall the dramatic diversity generated in cells of the adaptive immune system early in and throughout development. B-cells along with T-cells (named for the organs in which they were first identified —B for the bursa of birds and T for the thymus) are the major players of the adaptive, specific immune system discussed in chapter 2. They mix and match their DNA to create a vast array of molecules that might one day recognize an invader. These cells are activated by macrophages and other parts of the early, fast-acting, nonspecific immune system. The immune system "learns" to differentiate self from nonself, eliminating immune cells that recognize self; that is, because the immune system only attacks foreign invaders, it must be able to learn which cells are from the same organism as it is, so that it does not attack itself.

Mice that lack an immune system, or T-cells, have impairments in their *spatial learning ability*, which can be restored if T-cells are added back. During immune-system development throughout the life of mice, microbes affect the very nature of the repertoire of potential invader-recognition molecules (antibodies) that B-cells produce.[27] Wesemann and colleagues made two striking discoveries: first, in mice B-cells can develop and diversify *in the gut*, not only in the bone marrow as previously thought; and second, diversification of these gut B-cells is directly affected by signals from the gut microbiome. In other words, the nature and diversity of the immune system in these mice depends not just on learning and changing in response to their own self-cells, but also in response to their "own" *microbial* inhabitants, their microbiome.[28]

The definition of "the environment" is expanding to include the microbiome and other components of the astounding ecosystem existing in each individual. The developing immune system also learns from the vital environmental influence of microorganisms which microorganisms it is "wiser" not to attack but instead to ignore, because either these organisms are or could be beneficial, or attacking them is not worth the cost of the resulting chronic inflammation or lost tissues. The greater the diversity of microbes and parasites the immune system experiences, the better.

Human immune and nervous systems are so integrally linked that, as microorganisms help shape one, they inevitably help shape the other —in the womb and after birth. If the developing immune and nervous systems do not learn about Old Friends, or if the developing immune system is hit by maternal or early childhood infection, the result may be long-term susceptibility to inflammatory disease.

Let's now cultivate the quality of our attention with the mindfulness of
   breathing, placing our attention with each incoming and outgoing
   breath . . .
To anchor your attention in the present moment with the breath, place
   the attention at the nostrils or the abdomen . . .
Gently, as a butterfly so softly alights on a flower, simply attend to the
   unfolding physical sensations of each incoming and outgoing breath.

Perhaps now Chuck's hypothesis does not sound so strange: that the resilience of many Tibetans might be due in part, not only to their meditation practices, but also to their exposure to diverse bacteria, worms, and other microorganisms in the rural, herding, and farming environments of their youth. Because according to Chuck, the Old Friends hypothesis predicts that inflammation should be better regulated in *developing* countries, even though this might seem paradoxical given the high rate of infection in poorer nations.

The apparent conundrum is resolved by research demonstrating that people in low-income countries who have been exposed to Old Friends have a *sharp and discrete* response to infection; that is, they have an initial strong inflammatory response, which is quickly turned off once the stressor is gone. Maybe, then, this feature is a biological marker for resilience. In contrast, in wealthier populations not exposed to Old Friends, people's immune systems are chronically "on" without any apparent cause, increasing their risk for the "new" diseases.[29] Things are out of balance; there is an inability to maintain homeostasis.

While the ocean may be still in its depths, waves of all kinds arise on the surface when causes are there, and then naturally subside . . .
When we rest our awareness in the present moment, relate to the unfolding thoughts, feelings, memories, emotions, and so forth as the waves or ripples of our own mind . . . as we witness them without identifying with them or judging them, of their own accord they will subside into the stillness of our mind . . .
Just rest the awareness in that stillness or space between one thought and the next.

Balance. This is key. It's not the presence of a stress response per se, but the ability to regulate that response, to maintain equanimity, that is important—it's only on when a stressor is present, and then it's off.

Can meditation help strengthen the ability to maintain or reestablish that balance? Maybe it's not surprising that so much of the *biology* of meditation is related to maintaining a state of balance and stability. After all, much of the *act* of meditation and its purpose is itself about balance and stability.

My everyday meditation practices also help me to cultivate patience, and because of that, I never feel like giving up during many of the hard times

I have had. My three years of study abroad at Emory University were exciting and frustrating; there were many difficulties and challenges. But I never felt like giving up even a single day during those difficult times. I believe this is because of my patience and the result of my daily meditation practices.

I simply want to say that I meditate because it helps me to feel good, keep in good spirits, and remain joyful. All these joys contribute so much to my mental and emotional well-being. My regular meditation practice allows me to develop a deep trust in myself.

Some people do not like meditation practice; they feel like they are doing nothing and wasting time by meditating, because they are looking for speedy results. It can take some time for one to see results. I would say all one needs is patience and consistency with meditation. The little effort one makes will bring high compensation.

Thinking of balance and stability, Lobsang's guiding language for CBCT (which has been interwoven throughout this chapter), for example, refers to the still, relaxed position of the body: sit in a balanced, relaxed, even position, "witnessing the sensations, thoughts, images, feelings . . . in the spaciousness of the mind," keeping emotional balance, without judging or becoming "overly involved or entangled"; "check for any places of tension and . . . allow these places to soften," resituating extremes into the "middle path" and "shifting inner perspectives to align with reality"; and "anchor your attention."

The question is, then, from a biological standpoint: *Balance of what?*

Most scientific research papers are cited a handful of times, as many as a third, not at all; however, Kevin Tracey's paper, "The Inflammatory Reflex," published in *Nature* in 2002, has been cited over 2,200 times.[30] This is probably because the paper, a synthesis of the work of many scientists, offers tantalizing hints that it is *inflammation control* specifically that is the point, the fulcrum upon which the triangle of stress-response conversants is balanced.

TRACEY DEFINES the inflammatory reflex as the unconscious response that occurs when the immune system discovers bodily damage or foreign invasion, then induces inflammation and at nearly the same time also induces molecules to turn off inflammation. We saw this kind of built-in feedback in our discussion of the cortisol stress response in Kerry Ressler's work in chapter 4. Again, the balance between turning inflammation on—and to just the right amount—and turning inflammation off quickly and effectively is crucial. Too little response can lead to infection and worse; too much response to chronic inflammatory diseases, the "new" diseases.

The immune system and the nervous system are in constant communication. They produce some of the very same molecular communicators and receptors. For example, those most famous of *nervous system* molecules—neurotransmitters—are both produced and responded to by *immune system* cells. Once stimulated, the inflammatory reflex sends information to the hypothalamus in the brain, which then immediately activates an anti-inflammatory response.

The hypothalamus, as well as being key in the inflammatory reflex, is also important in another part of the nervous system stress response, the HPA axis, as we saw in chapter 4 in exploring resilience. This response, slower than the inflammatory reflex, involves a cycle of hormones. The first is released by the hypothalamus, and it eventually leads to the release of other stress hormones, notably cortisol (often measured as a marker of stress response), that circulate in the blood and stimulate energy production for stress response.

The inflammatory reflex is a process of the autonomic nervous system, so called because it operates largely (but, importantly as we shall see, not entirely) unconsciously, connecting the brain and vital organs to control breathing, heart rate, blood vessel dilation, and digestion.

As we continue to rest our awareness in the present moment, witnessing the inner reality as it unfolds, we may notice certain patterns of thoughts, concerns, or emotions that replay themselves frequently . . . Without judging or reacting or identifying with them, let's simply make note of them and allow them to subside in their own time, as they will, back into the stillness of our own mind.

Since chronic or excessive inflammation—perhaps because of a lack of exposure to Old Friends or overexposure to trauma—is common in many diseases, especially interesting are processes or activities that *decrease* inflammation.

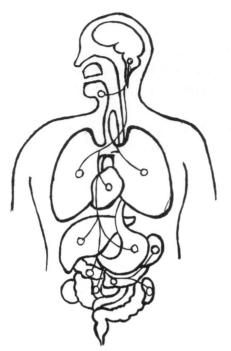

FIGURE 7.1 The vagus nerve's many contacts (indicated with circles) throughout the human body.

The vagus nerve is a major player in the inflammatory reflex of the autonomic nervous system; stimulation of it decreases inflammation. This nerve extends between many parts of the brain and all the major organs (in figure 7.1, small circles represent connections to organs). We first heard about the famous vagus nerve way back in chapter 1 in talking about its role in speech.

The organs talk to the vagus nerve and the brain (through *afferent* fibers), and the brain and vagus nerve talk to the organs (through *efferent* fibers). The vagus nerve helps monitor and regulate the body's internal state—for example, heart rate and blood pressure, which have a strong role in basic fight-or-flight stress responses, digestion, and temperature control. Darwin noted early on that the vagus nerve connects emotion to heart rate, stating in his book *The Expression of Emotions in Man and Animals* that "when the mind is strongly excited," heart rate increases.[31] Research is beginning to show correlations among vagus nerve activity, stress hormone levels, the immune system, and their potential role in gut dysfunctions such as Crohn's disease and irritable bowel syndrome.[32]

As we have seen in other situations, building evolutionary theories based on observation of contemporary organisms can give insight into complex processes. One such theory identifies three levels of the vagus nerve across evolutionary time. This is based on observing three different structures of the vagus nerve in organisms living today. At the first and most primitive level, in response to threat, the vagus nerve helps turn down metabolic, energy-producing activity in the organisms. The resulting behavior is immobilization in response to stress.

The second level of vagus nerve evolution allows an increase in energy production backup to prepare for "fight or flight." And the last evolutionary level of the vagus nerve, present only in humans and other mammals, allows fine and rapid control of heart rate. From such refined heart control comes the capacity to establish a "calmer state" helpful for growth and repair, as well as for refined interaction with the environment. This environment includes, importantly, other members of the species: the community.

Therefore, the theory goes, how organisms respond to stress and danger is directly related to their potential for social engagement, and the vagus nerve helps provide "the neurophysiological substrates for the emotional experiences and affective processes that are major components of social behavior."[33]

Greater and more refined regulation of the heart and increased communication with the brain occurred in evolution at around the same time as enhanced regulation and control of facial and vocal expression. Together these traits led to the potential for greater social engagement.

On the flip side, these evolutionarily new systems and possibilities involved in regulating fight or flight might cause damage if in chronic mode. Psychiatric disorders such as autism and depression often include symptoms corresponding to such predicted connections among social engagement, facial expression, and heart regulation by the vagus nerve: failure to meet gaze, overall decrease in facial affect, problems in chewing, as well as cardiovascular and digestive problems.[34]

Let's take a moment to look deeply within our own current experiences, reflecting on the factors that are troubling or unsettling us, thus undermining our happiness and contributing to the very distress and suffering we so wish to avoid.

The vagus nerve talks to the immune system via macrophages.[35] Macrophages are the cellular meeting place of brain and body, nerves and immunity. They are the Pac-men of the innate, first-response immune system, scanning continually for foreign invaders. Once they recognize an invader, macrophages attack and digest them and release molecules that cause inflammation. Macrophages are also important in releasing molecules that *decrease* inflammation. This makes sense in that inflammation is good in quick and small doses, but problematic if chronic; so if the very same cells that *induce* inflammation are also involved in *inhibiting* it, process efficiency is improved.

There is a strong likelihood that the third side of the stress-conversation triangle, the microbiome, is also either directly or indirectly affecting the behavior of macrophages. Microbiome bacteria have already been shown to increase the ability of one of macrophage's closest cousins (cells called neutrophils) to fight pathogens.[36]

Inflammation-causing molecules released by macrophages signal the brain through the afferent vagus nerve. The brain then interprets this information and via the efferent vagus nerve releases the neurotransmitter acetylcholine. Acetylcholine binds to macrophages, and they cease production of inflammatory molecules. This neural inflammatory off-on reflex is extremely fast—as compared to hormonal stress regulation—and can be localized to the site of inflammation or spread systemically if necessary.[37]

Having identified some of these factors or issues with which we struggle, let's take a few moments to reflect on them from a broader perspective . . .

Attuning to the universality of these events—gain and loss, success and failure, and even life and death—see if you can embrace this reality of change and impermanence as norms, not exceptions, as part of our shared human condition . . .

Once you have embraced this reality with some conviction, take a few moments to let this sink in more deeply.

If any part of the triangle of stress conversation is knocked out of whack —whether it be by physical or mental stress—inflammation may well result. According to Chuck, coevolution with pathogens and other threats, including psychological and social ones, left humans with an "inflammatory bias." In the modern world—short on Old Friends to check this inflammatory bias and rife with stressors real and imagined —humans' baseline stress, and thus inflammation, is shifted up chronically above zero, so that now the threshold for disease-causing inflammation is lowered. At the psychological level, for example, a result of this inflammatory bias then is "vulnerability to behavioral disorders that are coupled to inflammation, including reduced exploratory behavior in the form of depression and hypervigilance in the form of anxiety."[38]

People use meditation as a therapy not only in Tibetan medicine, but in some other traditions as well. Much of the wisdom, knowledge, and activities of the Tibetan culture in general is tightly tied to religion and ritual practices. Many Tibetans meditate as part of their spiritual practice, but not as a therapy, while others use it as a therapy, but not as part of spiritual practice.

I see close links among Tibetan medicine, ritual practice, and meditation practice. From the religious point of view, meditation is one of the best exercises to get rid of suffering and to attain enlightenment. In Tibetan medicine, meditation is a therapy associated with mental illness; it is helpful to reduce negative emotions. The more one is involved in meditation, the greater the benefits. You feel good. And regular meditation can then keep the state of mind more stable. If the mind is stable, that means you are in good health.

Tibetans commonly say, "Meditation's psychological benefits equal health benefits." Tibetan doctors and Buddhist practitioners consider the relationship between mind and body most important for well-being. Many diseases are caused by the imbalance of the three humors (wind, bile, and phlegm) and disturbances of the mind. Traditionally, when Tibetans have a problem, they seek different expertise to diagnose their diseases. They ask the help of physicians, ritual performers, meditators, and diviners. After examination of the patient, these practitioners may offer medicine or prescribe some other activities for the patient's well-being, such as healing rituals, meditation, prostration, and lifestyle changes. The majority of diseases are related to lifestyle.

These ideas predict that versions of genes which make humans susceptible to depression remain in the population because of their role in protective behavioral and immune responses to pathogens. That is, scientists know that versions of genes that stay in the population over time even though they appear "bad" are, for the very reason that they are remaining in the population, probably providing some potential "good." The "bad" here is depression and anxiety, while the "good" flip side is hypervigilance and reduced exploratory behaviors in times of stress. Consistent with this, researchers find associations between particular genetic changes in important inflammatory molecules and both increased risk of depression and decreased responsiveness to antidepressants.[39]

Let this insight guide you toward deeper conviction in the potential
to shape your inner life to emerge from these causes of suffering . . .
Take a few moments to immerse your heart and mind in this conviction,
perhaps visualizing this as a soothing light at the center of your chest,
expanding with each in-breath to fill completely your whole being.

Humans do, however, have conscious control, and thus can have substantial positive effect on the balance of the triangle of stress conversation. Meditation is an example of such conscious control. It may, for example, work in part through controlled breathing, which would activate the vagus nerve, which would decrease inflammation.

Breathing is absolutely necessary for survival. So it makes sense that it normally occurs unconsciously, is carefully monitored and regulated, and "has priority" in the brain. At the same time, it is perhaps the easiest such biological phenomenon to *consciously* alter. Indeed, ritualized breathing is central to CBCT and numerous other such practices—many of which also involve chanting and singing—across diverse cultures.[40] Information from the respiratory and cardiovascular system is constantly passed to the brain via the vagus nerve. Some of the positive effects on human health—on the immune, nervous, cardiovascular, and respiratory systems—provided by meditation may well be through breathing and its known vagus nerve connections.[41]

There is clearly more, though, to the positive effects of CBCT and other types of meditation than simply deep breathing. The effects probably manifest through breathing in combination with a number of other phenomena.

Think of meditation, then, as humans' active, conscious engagement in the three-way conversation that typically goes on within us unconsciously and continuously.

As we conclude this session, let's take a moment to dedicate our practice to the benefit of all sentient beings . . . and let's conclude by setting an intention to extend the insights that we have cultivated in this session to our everyday life.

THE YOUNG PEOPLE come in one at a time, ninety minutes before testing. They are all first-year college students, medically healthy. They don't know what's coming. Some have been practicing CBCT for six weeks on their own and with a group; others, matched controls, rather than meditate, spent time in health education classes that included discussions on stress management, community building exercises, and an assigned short research paper.

After being connected to monitors, the participants enter a room with three judges they have never seen before sitting behind a table. The judges look serious, a video camera looms. "It's a job interview," the participants are told. "Take this paper and pen. You have five minutes to prepare and then five minutes to convince us you are the one for the job." The judges look on impassively. Stress builds.

Just before the interview, the participants' notes are suddenly taken away. "Stand and talk." Still the judges are impassive. If the participants do not talk for the full five minutes, they are told to continue until they do so.

Finally, when finished, the participants are asked to count backward by thirteens from 1022; and if they make a mistake, they have to start over.

This is the Trier Social Stress Test, a controlled, replicable procedure for testing stress effects. Lobsang and Chuck, immersed in the biology of mind and body discussed here, decided with their colleagues to assess the effect of CBCT in several ways. Throughout the Trier Test, blood samples were taken to measure levels of the stress hormone cortisol as a marker of stress and of IL-6 as a marker for activation of the immune system and inflammation. Chuck and Lobsang also used a written psychological test before and after this stress test to measure the mood of participants.[42]

They started with healthy volunteers with no history of mental illness; so they were interested in effects on stress and immune response in relation to *prevention*.

Intriguingly, within the meditation group, those who spent more time practicing meditation had a significantly better immune response (lower IL-6 produced) than those who had practiced less. And there were trends, although not statistically significant, showing that more meditation correlated to a better mood in the meditation group.[43]

What if the improved immune response in meditators who practiced more CBCT was observed simply because people who already had a stronger stress response before the study were more likely to practice more CBCT? Lobsang and Chuck demonstrated with their colleagues that this was *not* the case. They proved this by exposing experiment participants to the stress test prior to CBCT training and then analyzing whether stress-response measures predicted the amount of meditation practice during training. They did not.[44]

Lobsang and Chuck's results make sense within the framework of the biology discussed in this chapter, the interaction with stress of immune, nervous, and other biologic systems. They are also typical of such pioneering studies in this area—promising, correlative, and in agreement with some past studies but not with others that use different types of meditation, practitioners, and controls.

Perhaps Lobsang and Chuck's results are showing that effects on the immune system or other biological effects are seen after only a certain threshold of CBCT practice time. This is in accord with other research on compassion-based meditation effects. In a different study, another meditative practice was shown to improve response to viral infection—in line with the CBCT IL-6 results, but the degree of immunity did not correlate with the amount of time spent meditating.[45]

If modern humans' baseline inflammation is high, perhaps meditation lowers that inflammation, improving stress response and making illness less likely in the future.

Can CBCT decrease inflammation and improve stress response therapeutically in those *already* suffering? Chuck, Lobsang, and several collaborators investigated this question in a group of adolescents in foster care who had experienced early life adversity, but at the time of the study were otherwise healthy. Half the group engaged in six weeks of CBCT training and practice and half the group did not. For ethical reasons, the young people were not subjected to a stress test, and an immune-system biomarker was taken from their saliva instead of from their blood before and after training.[46]

Again, *within* the meditating group, those who reported meditating more had less inflammatory biomarker, that is, they had a better immune response than those who meditated less. Just practicing CBCT during class time does not appear to have an effect, but how engaged participants are, how much time they spend with it on their own, makes a difference at the molecular level. In this study, the difference did not correlate to a difference in self-reported behavioral effects, but it might have if the study had continued over a longer period of time than six weeks.[47]

As discussed earlier, depression and anxiety are correlated with higher levels of inflammatory biomarkers, and interventions that decrease inflammation have antidepressive and antianxiety properties; so perhaps the foster-care children by practicing more CBCT are decreasing inflammation resulting from their adverse early-life experiences, which will then also improve their future stress response.

Recall the discussion in chapter 4 on how epigenetic changes link genes and experience. If meditation has the kind of profound effects on mind and body implied, epigenetic changes should occur in genes related to the biological processes involved, especially in long-term meditators. Indeed, as compared to controls, genes involved in general epigenetic regulation are turned down—that is, they are expressed at lower levels—in long-term meditators, meaning that meditation is correlated with broad changes in gene expression. In addition, a specific gene, known to be involved in increased inflammation, was also epigenetically turned down in the meditators.[48] Note that these are more than likely short-term epigenetic changes, or at the most, long-term within one individual, and not the kind of transgenerational epigenetic effects we also discussed in chapter 4.

In another striking link between meditation and molecular change, a large interdisciplinary team, similar to the one Chuck and Lobsang created, demonstrated significant connections among meditation, psychological health, and levels of the enzyme telomerase in immune cells in the blood.[49] Telomerase restores the ends of chromosomes after cell division; therefore, its activity serves as a marker for cellular longevity. The longer the chromosomal ends, the younger the cell. The greater the *perceived* stress of an individual, the shorter the telomeres, so that a significant disparity can exist between the chronological age (how long one has been alive) and the biological age (the length of one's telomeres) of one's cells.[50] Jacobs and collaborators showed that long-term meditators were in better psychological health and had more telomerase activity (and by implication longer, "younger" telomeres) than controls.[51]

Another question yet to be examined is whether meditation alters the microbiome. Our discussion so far would predict that it might well do so.

CHUCK AND LOBSANG, working with others including B. Alan Wallace (who collaborated in the telomerase study as well), also examined the impact of CBCT directly on the brain. It would be expected, of course, that all these molecular changes in inflammatory and nervous-system molecules affect the brain in significant ways. The brain, after all, is the primary synthesizer and monitor of all conversations.

If CBCT has effects on emotion regulation, one might see effects in the amygdala. Indeed, when Chuck, Lobsang, and their colleagues compare participants' depression scores and fMRIs of their amygdalas while participants observe images of suffering before and after CBCT, they find an interesting relationship. The stronger the amygdala response of the meditators who observe suffering, the greater their decrease in depression score. This effect could be because of an increased capacity for compassion, without corresponding empathic fatigue, from practicing CBCT.[52]

Other research of Chuck and Lobsang's both demonstrates CBCT effects on the brain and integrates its effects on inflammation with those on compassion. They demonstrated that CBCT improves practitioners' empathy and increases the excitation of empathy-associated parts of their brains to a greater extent than controls.[53]

Here, as they point out, Chuck and Lobsang's results are very similar to those induced by giving people oxytocin. Oxytocin is another hormone produced in the hypothalamus we mentioned previously; it is important in pair bonding, childbirth, and probably social recognition and altruism as well. It would make sense, but has yet to be shown, that CBCT also works by affecting the oxytocin system.[54]

In sum, Lobsang and Chuck's research, together with that of many others, strongly suggests the Dalai Lama is right, that this Eastward journeys narrative might really have something to offer: Western science strategically integrated with elements of ancient meditative practices can make a difference in people's lives and decrease suffering. It is also clear that the research into mechanisms and a true scientific grasp, if such a thing is possible, of the effects of meditation will require much more research. Consider that most of the scientific information—irrespective of meditation and its effects—discussed in this chapter has been uncovered only in the last decade or less.

A LOGICAL QUESTION that follows from Lobsang and Chuck's research: if contemplative practices can help improve lives and decrease suffering, if they can help students learn and teachers teach, if they can help deal with challenging and stressful emotional situations, why should they not be part of the education and experiences of physicians and other health care providers, those facilitating this healing? The question has far-reaching implications and applications, yet has been asked only recently and then only rarely.

When they enter medical school, students have mental health profiles similar to their peers doing other things; when they leave, their chances of being depressed or burned out are much greater. Physicians have a much greater risk of suicide—130 percent greater for females —than the population as a whole. All this despite their obvious easy access to health care. In one study, only a quarter of physicians who thought of themselves as depressed considered getting assistance, and only 2 percent actually sought help.[55]

Throughout this book, we explore how to most effectively bring what is learned in the Emory-Tibet Science Initiative into the *science* lab and classroom. As important are the classrooms and other learning environments of future and current health care providers. If a major goal is to heal and decrease suffering, the healers themselves ought not be forgotten.

Lobsang has also recently begun teaching CBCT to Emory medical students. The limited research on such work in this population suggests it is beneficial. Results of an analysis of several studies from around the world that together reached over a thousand physicians and medical students with various forms of mindfulness and stress-reduction interventions show significant decreases in anxiety and burnout.[56]

It is understandable, but ironic, that the measures of success of such approaches (and perhaps implying that the primary reason they are instituted in the first place) are very behavioral and biologically based and emphasize the *removal of the negative*—less burnout and anxiety. Such measures of success are, after all, strikingly indicative of the scientific worldview in which the researchers are operating in the first place; in the West health is often considered the absence of disease, rather than that plus mental and physical flourishing.

While fewer exhausted and stressed health care workers are obviously a desirable outcome, it is the *positive* effects of contemplative education, as suggested by Barbara McClintock and Arthur Zajonc in the previous chapter, that are at least as valuable. Such educational practices as Lobsang's CBCT should result in *better physicians and scientists*.

Better how?

Rodney Dietert, who has developed a rigorous contemplative education for Cornell undergraduates interested in biomedicine, elegantly addresses this question from the perspective of the skills a physician or scientist should ideally have. Typically emphasized in science education, as noted, are content knowledge, methodologies, and becoming a focused, critical thinker. But what about flexibility, innovation, and an ability to buck the trends and envision new paths? Dietert argues this is where contemplative practices and their associated "first-person" nature can be especially useful.[57]

An essential component of contemplative practices such as CBCT is the capacity to creatively take on different and multiple perspectives—a capacity extremely helpful for addressing problems such as those faced routinely by physicians. Contemplative practices teach self-awareness and the ability to move among perspectives (say from "scientific thinking" to emotional or more holistic and back again); in CBCT, an early step is to learn compassion for oneself. This requires knowing oneself. Clearly, self-aware, compassionate physicians are desirable also.

Dietert describes an extensive contemplative curriculum that moves well beyond meditation to engage in music, dance, visualization, journaling, improvisation, and attention to a prepared personal space, among other techniques—many of which have independently been proven effective in other areas of education. Most of Dietert's evidence —so far, based primarily on anecdotes, student evaluations, and student comments (so here, too, more research is needed)—suggests such a curriculum changes the way students approach problems in science and medicine and even how they go about choosing careers.

This is just the tip of the iceberg. More research results are on the way, judging from the dramatic growth in interest in contemplative studies and pedagogies and the resulting hundreds of health care practitioners, educators, and scientists, across innumerable disciplines, who are filling conferences on such topics these days.

This chapter began with Lobsang sitting next to Konchok in the front row of Emory's graduation ceremony. Konchok and his fellow monks had finished five years of science study in Dharamsala and then were honored for completing three more at Emory. They have become intellectual brothers with me and Lobsang. They have become bridges between two cultures and worldviews; they have returned to their monasteries and communities to establish science centers and help teach science throughout the year and with American scientists in the summers. With them and with Lobsang and this project, the whole ecology of our narrative is shifting—education, practice, modes of thinking, research, health, healing, and suffering.

I am myself a bridge between the world and culture of my people and the modern West. I see it in my father, and I see it in my experiences with our project. My father is a religious divination practitioner and traditional medicine practitioner. I would not be wrong if I said my father is a multitalented person. He learned religious and divination practice from a renowned teacher and traditional medicine practice from his dad, my grandfather. We come from a long lineage of traditional medicine practitioners. My eldest brother also has practiced traditional medicine, and now my youngest sister is training in this area.

All my life, I have not been able to spend much time with my father and other family members, because I left my home at an early age to become a monk, but I know my father treats his patients well. His first priority is examining the patient very carefully to find a diagnosis. He diagnoses diseases using simple methods: reading the pulse, checking eye and tongue color, and taking body temperature. Through these methods he can determine the causes of illnesses and diagnose problems of the inner organs.

Sometimes, my father also examines patient urine. I know a little bit about the methods modern doctors use to test urine and blood, because I myself have had these tests performed on me by Western doctors. My father's testing of urine is very different. Western doctors need special labs and equipment to test blood and urine, but my father does not use any tools or machines. Western doctors look at sugar and hemoglobin levels, chemicals, and the number of red and white blood cells, but my father uses the nature and intensity of urine smell and color to help in his diagnoses.

Mostly my father gives herbal medicine powders to his patients, but there are many problems my father does not have the tools to treat. If someone breaks a bone or injures a joint, he attempts to heal them by moxibustion, which is burning special herbs on the injury site. My father does this both for humans and domesticated animals with similar injuries. We also believe some illnesses are caused by *nagas* (serpent spirits). For these, my father performs ritual activities that I am sure would not be taken seriously by Western doctors. Of course, some illnesses or problems are not cured by any of my father's methods.

In their training, traditional medicine practitioners focus more on the practical and applied, rather than memorizing texts and information. They go into the meadows and other special areas at different elevations to collect medicinal plants as well as special stones that are heated to apply to painful sites on the body. My father and other practitioners learn to recognize healing herbs by their color, taste, shape, and the altitude at which they grow. They know the ingredients of each plant or herb and which are helpful to cure diseases.

Traditional medicines are often very helpful and have very few side effects. However, in some emergency cases—for example, when my mom was gored by a yak—traditional medicine is not very helpful. In such situations, Western medicine and technology are more powerful. If modern technology had been available in my village, my mother would still be alive. I still miss her.

I believe that for short-term problems, Western medicine and technology are best, and for long-term problems, traditional medicine is best. My father is very proud of me and of how I am learning about both traditions, helping to bring them together, and creating new ideas and thoughts that change how some people think and act—and that perhaps might even save some moms in the future.

CHAPTER EIGHT

Beyond Science and Religion

Q: Out of all the things you have done,
what is the most memorable?

A: I think one of them I did [in the] last thirty years or so
[is the] more serious sort of mutual learning with scientists . . .
so [in] our monastic institutions in India . . . science education
has been incorporated formally into the curriculum itself and
is now part of one of the required examination subjects . . .
now in [the] modern world science is very important,
so therefore [it] must be included. This is something
that has been an important achievement.

THE DALAI LAMA, BOSTON, 2014

e all stand because the Dalai Lama, in his basic black shoes and signature flowing maroon robes fringed in yellow, has left his residence and starts to head our way across the courtyard. But there is a long line of people who need blessing—newly married, newly born, rich, poor, white, Indian, Tibetan. He blesses them all. We sit and wait.

Then he finally starts our way again, but several people slip in between us for a photo or with things official for him to sign. Just before he gets to our door, a recent escapee from Tibet appears—a monk who snuck out of a prison in China and has been on the run for a year with a million-dollar bounty on his head. He and the Dalai Lama meet for a while.

Finally, there he is.

Konchok and I have been collecting and refining questions and getting our script together for months by e-mail, and in person the days before. Konchok is sweating like a monk about to talk to his supreme spiritual leader; apparently, in some situations no amount of meditation keeps you entirely calm.

When we sit down with the Dalai Lama, we are at a vital turning point in our project. We have just completed the five-year pilot in Dharamsala and within a week will start teaching the first classes of monks as part of their regular six-year curriculum in biology, neuroscience, and physics in the major monastic universities of Sera, Gaden, and Drepung Loseling.

We are eager to know the Dalai Lama's thoughts on the project, as well as what he thinks its impact has been and could yet be on Tibetan Buddhism specifically and on society as a whole.

He talks primarily in English, but often turns to his translator for particular words, usually recalling the words himself before the translator has a chance to respond, and then continuing. A few times he talks for several minutes in Tibetan, and then the translator fills us in.

The Dalai Lama tells the following story in response to one of our questions about the once and future relationship between science and religion. The story calls to mind the role China has played in the Eastward journeys narrative generally and in the American-Tibet story specifically.

In the 1950s, before escaping from China and into exile in India, the Dalai Lama met several times with Mao Zedong, the leader of the Chinese communist party. Oddly and ironically—given the near- and long-term future of the relationship between China and Tibet—the two men became very close: "He considered me as his son; I considered him as my father—very intimately."

In Beijing (known in the West as Peking at the time), the last time the Dalai Lama was there visiting, Mao unexpectedly and suddenly asked the Dalai Lama to come to his offices. No official translators were even present, and they had to ask one of Mao's bodyguards who happened to know both Chinese and Tibetan to translate.

The Dalai Lama sat at one end of a bench and Mao at the other, while Mao gave him "great advice" about how to lead people. After a while, Mao slid close and told the Dalai Lama, "You have a very scientific mind. Religion is opium."

The Dalai Lama says it was clear Mao thought that science and religion are very different, unrelated, and that religion is at best a tool of exploitation. "What would Mao think now in the twenty-first century when so much [is] being learned from science about religion and ethics and peace of mind in collaborative projects like ours?" the Dalai Lama wonders, his contagious laugh lifting the room's mood. "What would he think now?!"

The story carried many messages. First, apparent enemies can respect and listen to each other. Who would have suspected, based on today's dismal Sino-Tibet relations, that the Dalai Lama and Mao Zedong were so close? Second, the Dalai Lama is saying, yes, religion can be and has been misused and exploitative. He noted, for example, how the upper castes in India had often told the lower castes that their station in society is a result of their karma, and thus the rich and powerful use religion to justify mistreatment and abuse.

Third, in our conversation, I get the sense that the Dalai Lama is hinting at an analogy between the relationship he had with Mao and the relationship science has had with religion—tensions and dangers, but also possibilities and promise. The Dalai Lama uses the story to demonstrate how religion can be a powerful force for good, especially in relation to how science and religion can complement and learn from and respect each other.

When I compare how I was before this project and how I am now, I realize that I have gained and discovered so much, not only in Western science, but also in understanding Tibetan philosophy. The translation from one to the other makes both stronger.

Since I started learning Western science, it has totally changed my wisdom. I have rethought how to explore my own philosophy, how to explain or define certain profound concepts, how, in life, philosophy tenets can be used by normal people, not only by monks.

Western science has caused me to rethink deeply the formation of the universe, cause and effect, and karma. Learning science not only helps in understanding philosophy, but it also provides knowledge that can help in our daily lives.

We have not only learned science, but also how to teach well, and organize and lead classes.

Since I started learning science, I have started to see my own culture from a new perspective. I have been studying Tibetan philosophy for many years, and we believe in reincarnation, cause and effect, and karma; but so far we do not have evidence such as fossil records to prove these ideas. But we do have some significant signs and personal experiences that all these things are real. Science has made me rethink all these concepts.

The American anthropologist Margaret Mead said, "The traveler who has once been from home is wiser than he who has never left his own doorstep."[1] If you are inside your house, then you cannot see very much; when you come out from the house, you will see more and farther. So if you study something else besides yourself and your own, it changes you and makes you wiser.

The main point the Dalai Lama makes as we talk is the somewhat radical idea from a religious leader that science and religion can reinforce, strengthen, and maybe even enlighten each other. Such thinking is in strong contrast to the common narrative in the West (and also in communist China, as the Dalai Lama notes) that it is science *versus* religion, that science somehow disproves religion, and that religion should distrust, even fear, science.

Here's how the Dalai Lama develops this vital idea in his books (notably, *Ethics for the New Millennium*[2] and *Beyond Religion*[3]), in his talks, and that morning as we sit as close to him as he had to Mao more than half a century before.

"Secular ethics is very, very relevant," he says. "Secular ethics must use scientific findings. Your project could be a model. In order to create [a] happy twenty-first century—that means no war, no violence, and particularly, reduced fear and more equal[ity]." We must look for commonalities. Religions can be very helpful here, as all religions share similar core values such as compassion, forgiveness, love, and tolerance. Now the Dalai Lama makes one of the key turns in his argument.

In much the same way, as we saw in chapter 7, that meditation as a method has been secularized and removed from the religion of Tibetan Buddhism to treat stress and disease, these shared core human values need be *extracted from their religious contexts*. The core values—the Dalai Lama emphasizes compassion and interdependence especially— are separate from but part of all religions. These values form what he calls a system of "secular ethics."

"The way to promote secular ethics is not through preaching, but through education. So here, science is very, very important," the Dalai Lama tells us emphatically.

"Secular" is used here, the Dalai Lama explains, as it is in the Indian constitution—not to suggest atheism or antireligion, as the word often does in the United States. India is a country of many religious people, but the nation intentionally created the world's most populous democracy to encompass and engage all religions.

As my colleague ethicist Paul Root Wolpe points out, the Dalai Lama's secular ethics fits into an approach Western ethicists call "virtue ethics."[4] Very briefly, virtue ethics is a system with moral characteristics, or virtues, at its heart. Aristotle's questions of how to best define and live the good life embody such an approach. All other behaviors and principles start with and arise from these questions and the answers to them. Other ethical frameworks are driven more by rules, duties, or consequences of actions. Although classical Western virtue ethics includes the Aristotelian concept of *eudaimonia* ("happiness" or "flourishing"), the Dalai Lama's ethical system is the first based explicitly on compassion.

The Dalai Lama says we are all part of a shared, interdependent humanity striving for happiness and the avoidance of suffering; this is our "biological reality as social animals." All conscious beings, in fact, strive for happiness and fulfillment, to avoid suffering. When we are emotionally happy, we are also physically healthier, and this may well be what religion and science are all about. We all desire mental and physical health and well-being and meaningful relationships. This health, inner resilience, and purpose thrive via community and social connection, both of which are rooted in compassion.

There is "instinctive" compassion, such as that between mother and child, and "extended compassion" to our close relatives and beyond to all other humans and all sentient beings. Extended compassion can be cultivated and enhanced.

Compassion arises from empathy. We feel others' suffering and strive to relieve it. Compassion should not imply weakness—punishment and justice, for example, can also arise from compassion. From broad, unbiased compassion comes patience, kindness, forgiveness, and self-discipline. That is, from empathy and compassion, the other key virtues of secular ethics arise.

The next step is discernment: how do we compassionately weigh the harms and benefits of given situations? We often feel compassion for both sides in an issue. How can compassion guide us here most effectively? This is where Western systems of ethics—those based on rational analysis, duties, and consequences—might step in and assist and complement.

Science, says the Dalai Lama, is extremely important in teaching, developing, and supporting secular ethics. The science/secular ethics/ religion interconnection was initially just an individual idea of his, and if he promoted it, he says, a few people may or may not have listened; but now, major institutions—Emory, Stanford, and the University of Wisconsin—have taken up, expanded, and explored ideas of integrating science, ethics, teaching, and research. Now the potential for change is truly significant.

As we have shown in this book, science is an important and powerful source of evidence that strikingly and, in a fashion complementary to philosophical and religious evidence, demonstrates the importance and value of secular ethics. Compassion and empathy have evolved in us like other *biological* traits. This means that like other complex traits: (a) most humans have the *capacity* to be compassionate and empathic, and (b) humans can work intentionally to *improve* these traits in themselves.

The Dalai Lama says, gesturing toward the outdoors and laughing, "Nowadays, it's spring, birds now are nesting—some small babies, chicks—so mother is very, very busy feeding them out of her sense of concern, compassion—not from religious faith. I never heard these birds at some sort of prayer meeting!"

SCIENCE REINFORCES in a different way things we may have already known in an ethical or religious or intuitive sense, as well as challenging old ideas (here the Dalai Lama mentions the Tibetan Buddhist Mt. Meru creation story) and adding new potential possibilities for collaborative exploration. A healthy mind leads to a healthy body; as elaborated in chapter 5, humans are very much interdependent on each other, and all life on our planet is interlinked and interdependent.

The Dalai Lama adds that psychologists have shown that the more altruistic you are, the calmer your mind and the more self-confident you tend to be; the more self-centered, the more fear and unhappiness you experience. Secular ethics can be integrated into science discussions and classrooms, as discussed in earlier chapters, just as science can be integrated into the discourse of philosophy, ethics, and religion.

In our discussions with him, the Dalai Lama outlines this approach; in so doing he moves the conversation a step away from religion. So now it is much easier—still challenging, but much easier. He moves it from science and religion to science and ethics. In fact, as we talk, the Dalai Lama laughs and says he will toss some water on me from my cup, because I keep saying "science and religion" rather than "science and ethics." This is classic Dalai Lama: one of the world's best-known spiritual leaders, surrounded by security and in deep philosophical conversation, threatening to throw water at his conversation partner.

Religions are, of course, not unimportant to secular ethics. Religions, which will never go away the Dalai Lama notes, include *methods*—rituals and established protocols—that should be used to teach secular ethics: compassion, forgiveness, empathy, and tolerance. His approach allows the Dalai Lama to emphasize two points: (1) People should use those traditions and rituals with which they are most comfortable to grow and learn; no one religion is better than another; and (2) Religious practitioners should study the rigorous underlying philosophy, core values, and intellectual bases of their religions; sometimes these get lost among the rituals, "temples," and traditions.

"I say, too much temples. Of what use?" He laughs. "More explanation about spirituality, about ancient Indian psychology, much more useful than just chanting few words without knowing the meaning!"

The Dalai Lama relates the story of a woman who attended a series of his talks in the Netherlands: "Right from [the] beginning, I respect other religions. Recently, I was in Europe, in Holland . . . when I take some questions, one lady mentioned to me, "This morning you mentioned, according to Buddhism, [there is] no creator. She found [the] concept of [a] creator very, very helpful. She feels difficulty. I told her different religious traditions have different methods. They all teach us to practice love, forgiveness, tolerance, contentment. Your case, concept of God, creator, more suitable. You should believe that. It is much better. I am Buddhist; [the] concept of no creator [is] suitable for me. For you, concept of creator much better. This is our private business.

No problem. So, you see, I do not want to show Buddhism is something best. I never say that. We have no intention of conversion, for propagating Buddhism. That is why right from beginning, [I said in] Buddhism and science dialogues, we should keep [this] in mind."

While an individual of any religion (or one without any religion at all) can certainly engage the Dalai Lama's ideas on science, religion, and ethics, it is clear that Buddhism, notably Tibetan Buddhism, is a special case. As the Dalai Lama relates, it is perhaps easier to come to his formulations starting from Tibetan Buddhism than from other religions (for example, those that believe in a creator); and yet, the conclusions should work and make sense for anyone. He goes to great lengths, as he says, to emphasize that his intent has never been to say Buddhism is a "better religion" than any other. He has no interest in proselytizing. Rather than converting, people should work to fully appreciate their own religion's rituals and traditions.

The Dalai Lama notes that Tibetan Buddhism can be divided into Buddhist philosophy, Buddhist religion, and Buddhist science. Although the lines sometimes blur, such divisions probably make it easier for Tibetan Buddhism—as opposed to, say, Christianity—to collaborate with science. The Dalai Lama tells us he sees our project as a collaboration between Buddhist *science* and Western science. He says when he gets together with scientists, he does not discuss topics such as karma and reincarnation, but only the health of the mind and body —an area of clearer overlap and commonality between Buddhism and Western science.

The Dalai Lama also speaks to us of other long-held knowledge that facilitates Tibetan Buddhism's engagement of Western science: the ancient Indian Buddhist psychology he refers to that was brought to Tibet centuries ago. Today, Tibetan Buddhism is the only type of Buddhism that maintains and engages this body of work that interweaves so well with the modern science of the mind.

Gesturing toward the golden Buddha serenely observing us as we talk, the Dalai Lama says in response to a question about whether he is taking any risks in bringing science into the monasteries: "I don't think [so] . . . Buddha himself made it very clear that his teaching should not be accepted out of faith, out of devotion, but rather out of thorough investigation and experiment." He says that when we began our project, some of those monastic leaders in south India (home to the monastic universities in which we teach) were a little hesitant to engage science and make it part of their regular curricula; but when he reminded them of Buddha's teaching, that hesitancy disappeared.

The Dalai Lama tells us that whenever he communicates with other Buddhists—be it in China, Laos, Burma, Vietnam, Indonesia, or India —he encourages them to be "twenty-first-century Buddhists" in the spirit of the Nalanda tradition of Buddhism. This tradition dates from the fifth century AD in Bihar, India, and includes many influential Buddhist thinkers and philosophers. A central tenet of the Nalanda tradition is to study and discuss *all* extant worldviews, to explore all existing knowledge in the world; and since science is clearly a vital worldview, it should be studied and engaged (Konchok alludes to this element of Tibetan Buddhism in chapter 5).

We ask if in his mind one can be both a good practitioner of Buddhism and a good scientist. His response resonates with what Dhondup, one of the monks in our project, said about studying science so that he can understand his Buddhism better. "Now," says the Dalai Lama, "[I am] a more *intelligent* practitioner. . . . Like I always telling," he says with a laugh, "intelligent! This has implication that I'm among these intelligent people!"

Have any specific elements of his personal religious practices changed since he began studying science? "That I don't know," he says, and thinks for a minute. He says he started his practice when he was fourteen or fifteen years old, well before he formally engaged in science. But after some reflection, he tells us that the Western description of quantum physics has given him greater insight into his meditation on emptiness. And science has enriched and complemented his understanding of the potential connections between karma and evolution.

As discussed in this book, participants in our project—monks, nuns, professors, and undergraduate students—have, perhaps inevitably, stumbled into potential overlapping areas beyond the Dalai Lama's focus on the health of mind and body, beyond "Buddhist science," and into Buddhist philosophy and religion. We have made unexpected connections and realized parallels between, say, epigenetics and karma. Such connections have not been forced, but have emerged naturally. Any implications of such parallels are, so far, more in the philosophical realm than the scientific and utilitarian, but they are nevertheless real and worth exploring further.

At the end of our interview, the Dalai Lama takes my hand and holds it for a long time. He keeps saying "Thank you" (*he's* thanking *us!?*) and that our project has the potential to shift the world. "All 7.2 billion people," he says, because we must be as one and educate each other across all boundaries; we "can make some sort of significant contribution for humanity as a whole, like that. Thank you."

ACKNOWLEDGMENTS

Writing this book took a community of scholars from around the world. It could never have happened without His Holiness the Dalai Lama, who challenged us to create the Emory-Tibet Science Initiative, and Geshe Lobsang Negi, the human and spiritual link between Tibetan Buddhism and Emory University, a former monk, a college professor, and a dear friend who is an inspiration to many. This book was itself reincarnated several times before seeing the full light of day. Amy Benson Brown was one of the first to believe; Michael Fisher, with one phone call, brought the book back to life the first time; Wendell Wallach, who based on a chance (or karmic) meeting put us in contact with Andrew Stuart, who together with Paul Starobin, catalyzed another incarnation. David Westmoreland, as always, provided wisdom and humor. Tim Harrison, a wise and kind friend, read countless drafts and was a calm and enlightening critic and conversation partner. Lisa Hoveland Eisen was and is a constant light. The love and insight and genes of Jackie and Gene Eisen lift us up. We thank the dozens of professors from Emory and around the world who volunteered their time to teach the monks and nuns. Stephen Hull's deft touch and insight energized and sharpened our work. Konchok gives special and deep thanks to the abbot of his monastery, His Holiness Lungtok Tenpai Nyima, the Thirty-Third Menri Trizin. And finally, the students: monks, nuns, Emory undergraduates, who all keep us thinking and inspire boundless hope for the future.

NOTES

PROLOGUE

1. Portions of this book's story were first published in Arri Eisen, "What Buddhist Monks Taught Me about Teaching Science," *Chronicle of Higher Education* 58, no. 13 (November 18, 2011): B20, http://chronicle.texterity.com/chronicle /20111118b?folio=B20#pg20.

CHAPTER 1. ARE BACTERIA SENTIENT?

1. Arri Eisen and David Westmoreland, *The Living Staircase* (Dubuque, IA: Kendall/Hunt, 1998).

2. Nikhil Malvankar, Joanne Lau, Kelly Nevin, Ashley Franks, Mark Tuominen, and Derek Lovley, "Electrical Conductivity in a Mixed-Species Biofilm," *Applied and Environmental Microbiology* 78 (2012): 5967–71.

3. Elaine Patterson, John Cryan, Gerald Fitzgerald, R. Paul Ross, Timothy Dinan, and Catherine Stanton, "Gut Microbiota, the Pharmabiotics They Produce and Host Health," *Proceedings of the Nutrition Society* 73 (2014): 477–89.

4. Michael Bailey and Christopher Coe, "Maternal Separation Disrupts the Integrity of the Intestinal Microflora in Infant Rhesus Monkeys," *Developmental Psychobiology* 35 (1999): 146–55.

5. Graham Rook, Charles Raison, and Christopher Lowry, "Microbiota, Immunoregulatory Old Friends and Psychiatric Disorders," in *Microbial Endocrinology: The Microbiota-Gut-Brain Axis in Health and Disease*, ed. Mark Lyte and John Cryan (New York: Springer, 2014).

6. Nicole Gerardo and Alex Wilson, "The Power of Paired Genomes," *Molecular Ecology* 20 (2011): 2038–40.

7. Margaret McFall-Ngai and Edward Ruby, "Symbiont Recognition and Subsequent Morphogenesis as Early Events in an Animal-Bacterial Mutualism," *Science* 254 (1991): 1491–94.

8. Lynn Margulis, "The Origin of Plant and Animal Cells," *American Scientist* 59 (1971): 230–35.

9. Eisen, "What Buddhist Monks Taught Me."

Epigraph sources: Kitta MacPherson, "The 'Sultan of Slime': Biologist Continues to Be Fascinated by Organisms after Nearly Seventy Years of Study," *News at Princeton*, January 21, 2010.

Charles Darwin, *The Descent of Man* (New York: American Dome Library, 1901), 179.

1. John Bonner, *The Social Amoebae: The Biology of Cellular Slime Molds* (Princeton: Princeton University Press, 2009).

2. Ana Paula Sousa, Alexandra Amaral, Marta Baptista, Renata Tavares, Pedro Caballero Campo, Pedro Caballero Peregrín, Albertina Freitas, Artur Paiva, Teresa Almeida-Santos, and João Ramalho-Santos, "Not All Sperm Are Equal: Functional Mitochondria Characterize a Subpopulation of Human Sperm with Better Fertilization Potential," *PLOS One* 6 (2011): e18112.

3. R. John Aitken, Jock Findlay, Karla Hutt, and Jeff Kerr, "Apoptosis in the Germ Line," Reproduction 1441 (2011): 139–50.

4. Trisha Gura, "Reproductive Biology: Fertile Mind," *Nature* 491 (2012): 319–20.

5. Aitken et al., "Apoptosis in the Germ Line."

6. Ibid.

7. Melissa Pepling and Allan Spradling, "Mouse Ovarian Germ Cell Cysts Undergo Programmed Breakdown to Form Primordial Follicles," *Developmental Biology* 234 (2001): 339–51.

8. Qin Yan, Jin Ping Liu, and David Li, "Apoptosis in Lens Development and Pathology," *Differentiation* 74 (2006): 195–211.

9. M. Suzanne and H. Steller, "Shaping Organisms with Apoptosis," *Cell Death and Differentiation* 20 (2013): 669–75.

10. Veronica Del Riccio, Minke Van Tuyl, and Martin Post, "Apoptosis in Lung Development and Neonatal Lung Injury," *Pediatric Research* 55 (2004): 183–89.

11. Deepak Raj, Douglas Brash, and Douglas Grossman, "Keratinocyte Apoptosis in Epidermal Development and Disease," *Journal of Investigative Dermatology* 126 (2006): 243–57.

12. Cheryl Gatto and Kendal Broadie, "Fragile X Mental Retardation Protein Is Required for Programmed Cell Death and Clearance of Developmentally-Transient Peptidergic Neurons," *Developmental Biology* 356 (2011): 291–307.

13. David Dupret, Annabelle Fabre, Màtè Dàniel Döbrössy, Aude Panatier, José Julio Rodríguez, Stéphanie Lamarque, Valerie Lemaire, Stephane H. R Oliet, Pier-Vincenzo Piazza, and Djoher Nora Abrous, "Spatial Learning Depends on Both the Addition and Removal of New Hippocampal Neurons," *PLOS Biology* 5 (2007): 1683–94.

14. Ibid.

15. Brian Dias and Kerry Ressler, "Parental Olfactory Experience Influences Behavior and Neural Structure in Subsequent Generations," *Nature Neuroscience* 17 (2014): 89–96.

16. Ibid.

17. Drew Westen, Pavel Blagov, Keith Harenski, Clint Kilts, and Stephan Hamann, "Neural Bases of Motivated Reasoning: An fMRI Study of Emotional Constraints on Partisan Political Judgment in the 2004 US Presidential Election," *Journal of Cognitive Neuroscience* 18 (2006): 1947–58.

18. Laura-Ann Petitto and Kevin Dunbar, "Educational Neuroscience: New Discoveries from Bilingual Brains, Scientific Brains, and the Educated Mind," *Mind, Brain, and Education* 3 (2009): 185–97.

19. Hyung Don Ryoo and Andreas Bergmann, "The Role of Apoptosis-Induced Proliferation for Regeneration and Cancer," *Cold Spring Harbor Perspectives in Biology* 4 (2012): a008797.

CHAPTER 3. HOW DID LIFE BEGIN?

Epigraph source: Theodosius Dobzhansky, "Nothing Makes Sense Except in Light of Evolution," *American Biology Teacher* 35 (1973): 125–29

1. Theodosius Dobzhansky, "Nothing Makes Sense Except in Light of Evolution," *American Biology Teacher* 35 (1973): 125–29.

2. Ibid.

3. A. G. Cairns-Smith, *Genetic Takeover and the Mineral Origins of Life* (Cambridge: Cambridge University Press, 1982).

4. Ibid.

5. Cynthia Beall, Gianpiero Cavalleri, Libin Deng, Robert Elston, Yang Gao, Jo Knight, Chaohua Li, Jian Chuan Li, Yu Liang, Mark McCormack, Hugh Montgomery, Hao Pan, Peter Robbins, Kevin Shianna, Siu Cheung Tam, Ngodrop Tsering, Krishna Verramah, Wei Wang, Puchung Wangdui, Michael Weale, Yaomin Xu, Zhe Xu, Ling Yang, M. Justin Zaman, Changqing Zeng, Li Zhang, Xianglong Zhang, Pingcuo Zhaxi, and Yong Tang Zheng, "Natural Selection of *EPAS1* (*HIF2*) Associated with Low Hemoglobin Concentration in Tibetan Highlanders," *Proceedings of the National Academy of Sciences USA* 107 (2010): 11459–64.

6. C. G. Julian, J. L. Hageman, M. J. Wilson, E. Vargas, and L. G. Moore, "Lowland Origin Women Raised at High Altitude Are Not Protected Against Lower Uteroplacental O2 Delivery During Pregnancy or Reduced Birth Weight," *American Journal of Human Biology* 4 (2011): 509–16.

7. Francisco Ayala, "Darwin's Greatest Discovery: Design Without Designer," *Proceedings of the National Academy of Sciences* 104 (2007): 8567–73.

8. Charles Darwin, "Tree of Life," American Museum of Natural History (website), www.amnh.org/exhibitions/darwin/the-idea-takes-shape/i-think, accessed September 27, 2016.

9. Adrian Desmond and James Moore, *Darwin's Sacred Cause: How a Hatred of Slavery Shaped Darwin's Views on Human Evolution* (Chicago: University of Chicago Press, 2009).

10. Ibid.

11. Ibid.

12. Beall et al., "Natural Selection of *EPAS1* (*HIF2*)."

13. His Holiness the Dalai Lama, *How to See Yourself as You Really Are* (New York: Atria Books, 2006).

14. Jonathan Weiner, *The Beak of the Finch* (New York: Vintage Books, 1995).

15. Cecilia Lai, Simon Fisher, Jane Hurst, Faraneh Vargha-Khadem, and Anthony Monaco, "A Forkhead-Domain Gene Is Mutated in a Severe Speech and Language Disorder," *Nature* 413 (2001): 519–23.

16. Ayala, "Darwin's Greatest Discovery."

17. Ibid.

18. Dan-Erik Nilsson, "Eye Evolution and Its Functional Basis," *Visual Neuroscience* 30 (2013): 5–20.

19. Ibid.

20. Ibid.

21. Ibid.

22. WGBH Educational Foundation and Clear Blue Sky Productions, "Evolution of the Eye" (2001), www.pbs.org/wgbh/evolution/library/01/1/l_011_01.html.

23. Nilsson, "Eye Evolution and Its Functional Basis."

24. Arri Eisen and David Westmoreland, "Teaching Science, with Faith in Mind," *Chronicle of Higher Education* 55 (2009), http://chronicle.com/article/Teaching-Science-With-Faith/31764/.

25. Cynthia Passmore and Jim Stewart, "A Modeling Approach to Teaching Evolutionary Biology in High Schools," *Journal of Research in Science Teaching* 39 (2002): 185–204.

CHAPTER 4. ALTITUDE AND ATTITUDE

1. Ann Masten, "Ordinary Magic: Resilience Processes in Development," *American Psychologist* 56 (2001): 227–38.

2. Elissa Epel, Elizabeth Blackburn, Jue Lin, Firdaus Dhabhar, Nancy Adler, Jason Morrow, and Richard Cawthorn, "Accelerated Telomere Shortening in Response to Life Stress," *Proceedings of the National Academy of Sciences USA* 101 (2004): 17312–15.

3. Bill Clinton, "Remarks Made by the President," June 26, 2000, White House Office of the Secretary, National Human Genome Research Institute, www.genome .gov/10001356.

4. Miki Bundo, Manabu Toyoshima, Yohei Okada, Wado Akamatsu, Junko Ueda, Taeko Nemoto-Miyauchi, Fumiko Sunaga, Michihiro Toritsuka, Daisuke Ikawa, Akiyoshi Kakita, Motoichiro Kato, Kiyoto Kasai, Toshifumi Kishimoto, Hiroyuki Nawa, Hideyuki Okano, Takeo Yoshikawa, Tadafumi Kato, Kazuya Iwa-moto, "Increased L1 Retrotransposition in the Neuronal Genome in Schizophre-nia," *Neuron* 21 (2014): 306–13.

5. Kyle R. Upton, Daniel J. Gerhardt, J. Samuel Jesuadian, Sandra R. Richard-son, Francisco J. Sanchez-Luque, Gabriela O. Bodea, Adam D. Ewing, Carmen Sal-vador-Palomeque, Marjo S. van der Knaap, Paul M. Brennan, Adeline Vanderver, and Geoffrey J. Faulkner, "Ubiquitous L1 Mosaicism in Hippocampal Neurons," *Cell* 161 (2015): 228–39.

6. Gunnar Kaati, Lars Bygren, and Soren Edvinsson, "Cardiovascular and Dia-betes Mortality Determined by Nutrition During Parents' and Grandparents' Slow Growth Period," *European Journal of Human Genetics* 10 (2002): 682–88.

7. Ian Weaver, Nadia Cervoni, Frances Champagne, Ana D'Alessio, Shakti Sharma, Jonathan Seckl, Sergiy Dymov, Moshe Szyf, and Michael Meaney, "Epi-genetic Programming by Maternal Behavior," *Nature Neuroscience* 7 (2004): 847–54.

8. Ibid.

9. Sumeet Sharma, Abigail Powers, Bekh Bradley, and Kerry J. Ressler, "Gene X Environment Determinants of Stress- and Anxiety-Related Disorders," *Annual Review of Psychology* 67 (2016): 239–61.

10. Brandon A. Kohrt, Carol M. Worthman, Ramesh P. Adhikari, Nagendra P. Luitel, Jesusa M. G. Arevalo, Jeffrey Ma, Heather McCreath, Teresa E. Seeman, Eileen M. Crimmins, and Steven W. Cole, "Psychological Resilience and the Gene Regulatory Impact of Posttraumatic Stress in Nepali Child Soldiers," *Proceedings of the National Academy of Sciences* 113 (2016): 8156–61.

11. Negar Fani, Tricia Z. King, Emily Reiser, Elisabeth B. Binder, Tanja Jova-novic, Bekh Bradley, and Kerry J. Ressler, "FKBP5 Genotype and Structural In-tegrity of the Posterior Cingulum," *Neuropsychopharmacology* 39 (2014): 1206–13.

12. Mario Fraga, Esteban Ballestar, Maria F. Paz, Santiago Roper, Fernando Se-tien, Maria L. Ballestar, Damia Heine-Suner, Juan C. Cigudosa, Miguel Urioste, Javier Benitez, Manuel Boix-Chornet, Abel Sanchez-Aguilera, Charlotte Ling, Emma Carlsson, Pernille Poulsen, Allan Vaag, Zarko Stephan, Tim D. Spector, Yue-Zhong Wu, Christoph Plass, and Manel Esteller, "Epigenetic Differences Arise during the Lifetime of Monozygotic Twins," *Proceedings of the National Acad-emy of Sciences USA* 102 (2005): 10604–609.

13. Arri Eisen and Junjian Huang, "Learning Science by Engaging Religion: A Novel Two-Course Approach for Biology Majors," *College Teaching* 62 (2014): 25–31.

14. Arieh Moussaieff, Neta Rimmerman, Tatiana Bregman, Alex Straiker, Christian Felder, Shai Shoham, Yoel Kashman, Susan Huang, Hyosang Lee, Esther Shohami, Ken Mackie, Michael J. Caterina, J. Michael Walker, Ester Fride, and Raphael Mechoulam, "Incensole Acetate, an Incense Component, Elicits Psychoactivity by Activating TRPV3 Channels in the Brain," *Journal of the Federation of American Societies for Experimental Biology* 22 (2008): 3024–34.

15. Eisen and Huang, "Learning Science by Engaging Religion."

16. Ibid.

17. Ibid.

CHAPTER 5. ECOLOGY AND KARMA

Epigraph sources: "The Sheltering Tree of Interdependence: Buddhist Monks' Reflections on Ecological Responsibility," poem for the occasion of the presentation by His Holiness the Dalai Lama of a statue of the Buddha to the people of India and to mark the opening of the International Conference on Ecological Responsibility: A Dialogue with Buddhism (1993), www.dalailama.com/messages /environment/buddhist-monks-reflections, accessed September 28, 2016.

"Ecological Buddhism: A Buddhist Response to Global Warming," editorial, www.ecobuddhism.org/wisdom/editorials/bcee, accessed September 28, 2016.

1. "Buddhism and the Climate—Energy Emergency," editorial, Ecological Buddhism: A Buddhist Response to Global Warming, www.ecobuddhism.org/wisdom /editorials/bcee, accessed September 28, 2016.

2. Ibid.

3. Daniel Miller, *Drokpa: Nomads of the Tibetan Plateau and Himalaya* (Gainesville, FL: Vajra Publications, 2008).

4. Ibid.

5. Ibid.

6. Gillian Tan, *Re-examining Human-Nonhuman Relations among Nomads of Eastern Tibet* (Geelong, Australia: Alfred Deakin Research Institute, Deakin University, 2012).

7. Lance Gunderson and C. S. Holling, eds., *Panarchy: Understanding Transformations in Human and Natural Systems* (Washington, DC: Island Press, 2002).

8. Brian Walker, C. S. Holling, Stephen R. Carpenter, Ann Kinzig, "Resilience, Adaptability and Transformability in Social-Ecological Systems," *Ecology and Society* 9, no. 2 (2004).

9. Marten Scheffer, Steve Carpenter, Jonathan Foley, Carl Folke, and Brian Walker, "Catastrophic Shifts in Ecosystems," *Nature* 413 (2001): 591–96.

10. Steven Southwick, George Bonanno, Ann Masten, Catherine Panter-Brick,

and Rachel Yehuda, "Resilience Definitions, Theory, and Challenges: Interdisciplinary Perspectives," *European Journal of Psycho-Traumatology* 5 (2104): 25338.

11. Ibid.

12. Ibid.

13. Sumeet Sharma, Abigail Powers, Bekh Bradley, and Kerry J. Ressler, "Gene X Environment Determinants of Stress- and Anxiety-Related Disorders," *Annual Review of Psychology* 67 (2016): 239–61.

14. Ingrid van de Leemput, Marieke Wichers, Angélique Cramer, Denny Borsboom, Francis Tuerlinckx, Peter Kuppens, Egbert van Nes, Wolfgang Viechtbauer, Erik Giltay, Steven Aggen, Catherine Derom, Nele Jacobs, Kenneth Kendler, Han van der Maas, Michael Neale, Frenk Peeters, Evert Thiery, Peter Zachar, and Marten Scheffer, "Critical Slowing Down as Early Warning for the Onset and Termination of Depression," *Proceedings of the National Academy of Sciences USA* 111 (2014): 87–92.

15. Lei Dai, Daan Vorselen, Kirill Korolev, and Jeff Gore, "Generic Indicators for Loss of Resilience Before a Tipping Point Leading to Population Collapse," *Science* 336 (2012): 1175–77.

16. van de Leemput et al., "Critical Slowing Down."

17. Leo Lahti, Jarkko Salojärvi, Anne Salonen, Marten Scheffer, and Willem de Vos, "Tipping Elements in the Human Intestinal Ecosystem," *Nature Communications* 5 (2014): 4344.

18. P. W. Anderson, "More Is Different: Broken Symmetry and the Nature of the Hierarchical Structure of Science," *Science* 177 (1972): 393–96.

19. Chinese Academy of Sciences, "Environmental Changes on the Tibetan Plateau: Evaluation and Prediction," *Bulletin of the Chinese Academy of Sciences* 29 (2015): 195–96.

20. Grant Lichtman, "Teaching 254 Kids a Year," *Learning Pond* (2013), www.grantlichtman.com/teaching-254-kids-a-year.

21. John Dewey, *Experience and Education* (Toronto: Collier-MacMillan Canada, 1938). International Centre for Educators' Learning Styles, "John Dewey's Philosophy of Experience and Education," https://eiclsresearch.wordpress.com/types-of-styles/teaching-styles/john-dewey/, accessed September 28, 2016.

22. Dewey, *Experience and Education.*

23. Ibid.

24. Ibid.

25. Urie Bronfenbrenner, "The Experimental Ecology of Education," *Educational Researcher* 5 (1976): 5–15.

26. Eisen and Huang, "Learning Science by Engaging Religion."

27. Carol Brewer and Diane Smith, eds., *Vision and Change in Undergraduate Biology Education: A Call to Action* (Washington, DC: American Association for the Advancement of Science, 2011), http://visionandchange.org/finalreport.

28. Ibid.

29. Arthur Zajonc, "Contemplative and Transformative Pedagogy," *Kosmos Journal* 5 (2006): 1–3.

CHAPTER 6. ARE HUMANS INHERENTLY GOOD?

Epigraph sources: His Holiness the Dalai Lama, www.facebook.com/DalaiLama, accessed December 29, 2014.

His Holiness the Dalai Lama and Howard Cutler, *The Art of Happiness: A Handbook for Living* (New York: Riverhead Books, 2009), 63.

1. Marc Iacoboni and Mirella Dapretto, "The Mirror Neuron System and the Consequences of Its Dysfunction," *Nature Reviews Neuroscience* 7 (2006): 942–51.

2. Cristina Gonzalez-Liencres, Georg Juckel, Cumhur Tas, Astrid Friebe, and Martin Brune, "Emotional Contagion in Mice: The Role of Familiarity," *Behavioural Brain Research* 263 (2014): 16–21.

3. V. S. Ramachandran, "Mirror Neurons and Imitation Learning as the Driving Force behind 'The Great Leap Forward' in Human Evolution," *Edge*, http://edge.org/3rd_culture/ramachandran/ramachandran_p1.html, accessed September 28, 2016.

4. Ronit Yoeli-Tlalim, "Tibetan 'Wind' and 'Wind' Illnesses: Towards a Multicultural Approach to Health and Illness," *Studies in the History and Philosophy of Biological and Biomedical Science* 41 (2010): 318–24.

5. His Holiness the Dalai Lama, *An Open Heart: Practicing Compassion in Everyday Life* (New York: Back Bay Books, 2001) 153–54.

6. W. A. Escobar, "Quantized Visual Awareness," *Frontiers in Psychology*, http://journal.frontiersin.org/article/10.3389/fpsyg.2013.00869/full, accessed September 28, 2016.

7. Philip Ball, "We Might Live in a Computer Program, but It May Not Matter," BBC, September 5, 2016, www.bbc.com/earth/story/20160901-we-might-live-in-a-computer-program-but-it-may-not-matter.

8. Curtis White, *The Science Delusion: Asking the Big Questions in a Culture of Easy Answers* (New York: Melville House, 2013).

9. His Holiness the Dalai Lama, *An Open Heart*, 153–54.

10. B. Alan Wallace, "A Science of Consciousness: Buddhism, the Modern West," *Pacific World: Journal of the Institute of Buddhist Studies* 4 (2002): 15–32.

11. *Charles Darwin, The Expression of the Emotions in Man and Animals* (London: John Murray, 1872).

12. Richard Moore, *Can I Give Him My Eyes?* (Ireland: Hachetter Books, 2009).

13. Gonzalez-Liencres et al., "Emotional Contagion in Mice."

14. His Holiness the Dalai Lama, "Compassion and the Individual," www.dalailama.com/messages/compassion, accessed September 28, 2016.

15. Ibid.

16. Frans de Waal, *The Age of Empathy: Nature's Lessons for a Kinder Society* (New York: Three Rivers Press, 2009).

17. Frans de Waal, "The Antiquity of Empathy," *Science* 336 (2012): 874–76.

18. Iain Douglas-Hamilton, Shivani Bhalla, George Wittemyer, and Fritx Vollrath, "Behavioural Reactions of Elephants towards a Dying and Deceased Matriarch," *Applied Animal Behaviour Science* 100 (2006): 87–102.

19. Daejong Jeon, Sangwoo Kim, Mattu Chetana, Daewoong Jo, H. Earl Ruley, Shih-Yao Lin, Dania Rabah, Jean-Pierre Kinet, and Hee-Sup Shin, "Observational Fear Learning Involves Affective Pain System and $Ca_v1.2Ca2+$ Channels in ACC," *Nature Neuroscience* 13 (2010): 482–88.

20. Ibid.

21. Inbal Ben-Ami Bartal, Jean Decety, and Peggy Mason, "Empathy and Pro-Social Behavior in Rats," *Science* 334 (2011): 1427–30.

22. Joshua Plotnik and Frans de Waal, "Asian Elephants (*Elephas maximus*) Reassure Others in Distress," *PeerJ* 2 (2014): e278.

23. Shinya Yamamoto, Tatyana Humle, and Masayuki Tanaka, "Chimpanzees' Flexible Targeted Helping Based on an Understanding of Conspecifics' Goals," *Proceedings of the National Academy of Sciences USA* 109 (2012): 3588–92.

24. de Waal, *Age of Empathy*.

25. Jennifer Mascaro, Thaddeus Pace, and Charles Raison, "Mind Your Hormones! The Endocrinology of Compassion," in *Compassion: Bridging Practice and Science*, ed. Tania Singer and Matthias Bolz (Munich: Max Planck Society, 2013), www.compassion-training.org, accessed September 29, 2016.

26. Ibid.

27. Iacoboni and Dapretto, "The Mirror Neuron System and the Consequences of Its Dysfunction." *Nature Reviews Neuroscience* 7 (2006): 942–51.

28. Jonathan Prather, Susan Peters, Stephen Nowicki, and Richard Mooney, "Precise Auditory-Vocal Mirroring in Neurons for Learned Vocal Communication," *Nature* 451 (2008): 305–10.

29. Chris Woolston, "Researcher Under Fire for *New Yorker* Epigenetics Article," *Nature* 533 (2016), www.nature.com/news/researcher-under-fire-for-new-yorker-epigenetics-article-1.19874.

30. Sally Satel and Scott Lilienfeld, *Brainwashed: The Seductive Appeal of Mindless Neuroscience* (New York: Basic Books, 2013).

31. Philippe Rochat, "Various Kinds of Empathy as Revealed by the Developing Child, Not the Monkey's Brain," *Behavioral and Brain Sciences* 25 (2002): 45–46.

32. Ibid.

33. Matthew Cohen, Deqiang Jing, Rui Yang, Nim Tottenham, Francis lee, and B. J. Casey, "Early-Life Stress Has Persistent Effects on Amygdala Function and Development in Mice and Humans," *Proceedings of the National Academy of Sciences USA* 110 (2013): 18274–78.

34. Mona Sobhani, Glenn Fox, Jonas Kaplan, and Lisa Aziz-Zadeh, "Interpersonal Liking Modulates Motor-Related Neural Regions," *PLOS One* 7 (2012): e46809.

35. Ruben Azevedo, Emiliano Macaluso, Alessio Avenanti, Valerio Santangelo, Valentina Cazzato, and Salvatore Maria Aglioti, "Their Pain Is Not Our Pain: Brain and Autonomic Correlates of Empathic Resonance with the Pain of Same and Different Race Individuals," *Human Brain Mapping* 34 (2013): 3168–81.

36. Steven Pinker, *The Better Angels of Our Nature: The Decline of Violence in History and Its Causes* (New York: Penguin Group, 2011).

37. Andrea Olsson, Jeffrey Ebert, Mahzarin Banaji, and Elizabeth Phelps, "The Role of Social Groups in the Persistence of Learned Fear," *Science* 309 (2005): 785–87.

38. His Holiness the Dalai Lama, *Beyond Religion: Ethics for a Whole World* (New York: Houghton Mifflin Harcourt, 2011).

39. Hooria Jazaieri, Kelly McGonigal, Thupten Jinpa, James Doty, James Gross, and Philippe Goldin, "A Randomized Controlled Trial of Compassion Cultivation Training: Effects on Mindfulness, Affect, and Emotion Regulation," *Motivation and Emotion* 38 (2013): 23–35.

40. Olga Klimecki, Susanne Leiberg, Claus Lamm, and Tania Singer, "Functional Neural Plasticity and Associated Changes in Positive Affect after Compassion Training," *Cerebral Cortex* 23 (2013): 1552–61.

41. Helen Weng, Andrew Fox, Alexander Shackman, Diane Stodola, Jessica Caldwell, Matthew Olson, Gregory Rogers, and Richard Davidson, "Compassion Training Alters Altruism and Neural Responses to Suffering," *Psychological Science* 24 (2013): 1171–80.

42. Jennifer Mascaro, James Rilling, Lobsang Negi, and Charles Raison, "Compassion Meditation Enhances Empathic Accuracy and Related Neural Activity," *Social Cognitive and Affective Neuroscience* 8 (2013): 48–55.

43. David Perlman, Tim Salomons, Richard Davidson, and Antoine Lutz, "Differential Effects on Pain Intensity and Unpleasantness of Two Meditation Practices," *Emotion* 10 (2010): 65–71.

44. Joseph Durlak, Roger Weissberg, Allison Dymnicki, Rebecca Taylor, and Kriston Schellinger, "The Impact of Enhancing Students' Social and Emotional Learning: A Meta-Analysis of School-Based Universal Interventions," *Child Development* 82 (2011): 405–32.

45. Evelyn Fox Keller, *A Feeling for the Organism: The Life and Work of Barbara McClintock* (New York: W. H. Freeman, 1983), 203–4.

46. Ibid.

47. Arthur Zajonc, "Cognitive-Affective Connections in Teaching and Learning: The Relationship between Love and Knowledge," *Journal of Cognitive Affective Learning* 3 (2006): 1–9.

CHAPTER 7. MEDITATION AND THE "NEW" DISEASES

1. Gaëlle Desbordes, Lobsang Negi, Thaddeus Pace, B. Alan Wallace, Charles L. Raison, and Eric Schwartz, "Effects of Mindful-Attention and Compassion Meditation Training on Amygdala Response to Emotional Stimuli in an Ordinary, Non-Meditative State," *Frontiers in Human Neuroscience* 6 (2012): 1–15.

2. Thaddeus Pace, Lobsang Negi, Daniel Adame, Steven Cole, Teresa Sivilli, Timothy Brown, Michael J. Issa, and Charles L. Raison, "Effect of Compassion Meditation on Neuroendocrine, Innate Immune and Behavioral Responses to Psychosocial Stress," *Psychoneuroendocrinology* 34 (2009): 87–98.

3. Anne Harrington, *The Cure Within: A History of Mind-Body Medicine* (New York: W. W. Norton, 2009).

4. Ibid.

5. Ibid.

6. Ibid.

7. Ibid.

8. Ibid.

9. Ibid.

10. Ibid.

11. Desbordes et al., "Effects of Mindful-Attention."

12. Andrew Vickers, Niraj Goyal, Robert Harland, and Rebecca Rees, "Do Certain Countries Produce Only Positive Results? A Systematic Review of Controlled Trials," *Controlled Clinical Trials* 19 (1998): 159–66.

13. David Colquhoun and Steven Novella, "Acupuncture Is Theatrical Placebo," *Anesthesia and Analgesia* 116 (2013): 1360–63.

14. Chris Streeter, Patricia Gerbarg, Robert Saper, Domenic Ciraulo, and Richard Brown, "Effects of Yoga on the Autonomic Nervous System, Gamma-Aminobutyric-Acid, and Allostasis in Epilepsy, Depression, and Post-Traumatic Stress Disorder," *Medical Hypotheses* 78 (2012): 571–79.

15. Rook, Raison, and Lowry, "Microbiota."

16. Charles Raison, "How Alzheimer's Disease Is Linked with Hygiene," October 6, 2013, PsychCongress Network, www.psychcongress.com/blog/how-alzheimer's-disease-linked-hygiene.

17. Elaine Hsiao, Sara McBride, Sophia Hsien, Gil Sharon, Embriette Hyde, Tyler McCue, Julian Codelli, Janet Chow, Sarah Reisman, Joseph Petrosino, Paul Patterson, and Sarkis Mazmanian, "The Microbiota Modulates Gut Physiology and Behavioral Abnormalities Associated with Autism," *Cell* 155 (2013): 1451–63.

18. Leo Lahti, Jarkko Salojarvi, Anne Salonen, Marten Scheffer, and Willem M. de Vos, "Tipping Elements in the Human Intestinal Ecosystem," *Nature Communications* 5 (2014): 4344.

19. Mohamed Donia, Peter Cimermancic, Christopher Schulze, Laura Wieland

Brown, John Martin, Makedonka Mitreva, Jon Clardy, Roger Linington, and Michael Fischbach, "A Systematic Analysis of Biosynthetic Gene Clusters in the Human Microbiome Reveals a Common Family of Antibiotics," *Cell* 158 (2014): 1402–14.

20. Michael Bailey and Christopher Coe, "Maternal Separation Disrupts the Integrity of the Intestinal Microflora in Infant Rhesus Monkeys," *Developmental Psychobiology* 35 (1999): 146–55.

21. Ibid.

22. Rook, Raison, and Lowry, "Microbiota."

23. Ibid.

24. Ibid.

25. Charles Raison and Andrew Miller, "Malaise, Melancholia, and Madness: The Evolutionary Legacy of the Inflammatory Bias," *Brain, Behavior, and Immunity* 31 (2014): 1–8.

26. Aswin Sekar, Allison Bialas, Heather de Rivera, Avery Davis, Timothy Hammond, Nolan Kamitaki, Katherine Tooley, Jessy Presumey, Matthew Baum, Vanessa Van Doren, Biulio Benovese, Samuel Rose, Robert Handsaker, Schizophrenia Working Group of the Psychiatric Genomics Consortium, Mark Daly, Michael Carroll, Beth Stevens, and Steven McCarroll, "Schizophrenia Risk from Complex Variation of Complement Component 4," *Nature* 530 (2016): 177–83.

27. Duane Wesemann, Andrew Portuguese, Robin Meyers, Michael Gallagher, Kendra Cluff-Jones, Jennifer Magee, Rohit Panchakshari, Scott Rodig, Thomas Kepler, and Frederick Alt, "Microbial Colonization Influences Early B-Lineage Development in the Gut Lamina Propria," *Nature* 501 (2013): 112–15.

28. Ibid.

29. Rook, Raison, and Lowry, "Microbiota."

30. Kevin Tracey, "The Inflammatory Reflex," *Nature* 420 (2002): 853–59.

31. Stephen Porges, "The Polyvagal Theory: New Insights into Adaptive Reactions of the Autonomic Nervous System," *Cleveland Clinic Journal of Medicine* 76 (2011): S86–S90.

32. Sonia Pellissier, Cecile Dantzer, Laurie Mondillon, Candice Trocme, Anne-Sophie Gauchez, Veronique Ducros, Nicolas Mathieu, Bertrand Toussanit, Alicia Fournier, Frederic Canini, and Bruno Bonaz, "Relationship between Vagal Tone, Cortisol, TNF-alpha, Epinephrine and Negative Affect in Crohn's Disease and Irritable Bowel Syndrome," *PLOS One* 9 (2014): e105328.

33. Porges, "The Polyvagal Theory."

34. Ibid.

35. Tracey, "The Inflammatory Reflex."

36. Thomas Clarke, Kimberly Davis, Elena Lysenko, Alice Zhou, Yimin Uy, and Jeffrey Weiser, "Recognition of Peptidoglycan from the Microbiota by Nod1 Enhances Systemic Innate Immunity," *Nature Medicine* 16 (2010): 228–31.

37. Tracey, "Inflammatory Reflex."

38. Raison and Miller, "Malaise, Melancholia, and Madness."

39. Rook, Raison, and Lowry, "Microbiota."

40. Streeter et al., "Effects of Yoga."

41. Julian Thayer, Adrian Loerbroks, and Esther Sternberg, "Inflammation and Cardiorespiratory Control: The Role of the Vagus Nerve," *Respiratory Physiology and Neurobiology* 178 (2011): 387–94.

42. Pace et al., "Effect of Compassion Meditation."

43. Ibid.

44. Thaddeus Pace, Lobsang Negi, Teresa Sivilli, Michael J. Issa, Steven Cole, Daniel Adame, and Charles L. Raison, "Innate Immune, Neuroendocrine and Behavioral Responses to Psychosocial Stress Do Not Predict Subsequent Compassion Meditation Practice Time," *Psychoneuroendocrinology* 35 (2010): 310–15.

45. Nani Morgan, Michael Irwin, Mei Chung, and Chenchen Wang, "The Effects of Mind-Body Therapies on the Immune System: Meta-Analysis," *PLOS One* 9 (2014): e100903.

46. Thaddeus W. W. Pace, Lobsang Tenzin Negi, Brooke Dodson-Lavelle, Brendan Ozawa-de Silva, Sheethal D. Reddy, Steven P. Cole, Andrea Danese, Linda W. Craighead, and Charles L. Raison, "Engagement with Cognitively-Based Compassion Training Is Associated with Reduced Salivary C-Reactive Protein from before to after Training in Foster Care Program Adolescents," *Psychoneuroendocrinology* 38 (2013): 294–99.

47. Ibid.

48. Perla Kaliman, Maria Jesus Alvarez-Lopez, Marta Cousin-Tomas, Melissa Rosenkranz, Antoine Lutz, and Richard Davidson, "Rapid Changes in Histone Deacetylases and Inflammatory Gene Expression in Expert Meditators," *Psychoneuroendocrinology* 40 (2014): 96–107.

49. Tonya L. Jacobs, Elissa S. Epel, Jue Lin, Elizabeth H. Blackburn, Owen M. Wolkowitz, David A. Bridwell, Anthony P. Zanesco, Stephen R. Aichele, Baljinder K. Sahdra, Katherine A. MacLean, Brandon G. King, Phillip R. Shaver, Erika L. Rosenberg, Emilio Ferrer, B. Alan Wallace, and Clifford D. Saron, "Intensive Meditation Training, Immune Cell Telomerase Activity, and Psychological Mediators," *Psychoneuroendocrinology* 36 (2011): 664–81.

50. Elissa Epel, Elizabeth Blackburn, Jue Lin, Firdaus Dhabhar, Nancy Adler, Jason Morrow, and Richard Cawthorn, "Accelerated Telomere Shortening in Response to Life Stress," *Proceedings of the National Academy of Sciences USA* 101 (2004): 17312–15.

51. Jacobs et al., "Intensive Meditation Training."

52. Desbordes et al., "Effects of Mindful-Attention."

53. Jennifer Mascaro, James Rilling, Lobsang Negi, and Charles Raison, "Pre-existing Brain Function Predicts Subsequent Practice of Mindfulness and Compassion Meditation," *NeuroImage* 69 (2013): 35–42.

54. Jennifer Mascaro, Thaddeus Pace, and Charles Raison, "Mind Your Hormones! The Endocrinology of Compassion," in *Compassion: Bridging Practice and Science*, ed. Tania Singer and Matthias Bolz (Munich: Max Planck Society, 2013), www.compassion-training.org, accessed September 30, 2016.

55. Michael Myers, Todd Watkins, and Gisele Microys, eds., *Canadian Medical Association Guide to Physician Health and Well-Being* (Ottawa, Canada: Canadian Medical Association, 2003).

56. Cheryl Regehr, Dylan Glancy, Annabel Pitts, and Vicki LeBlanc, "Interventions to Reduce the Consequences of Stress in Physicians: A Review and Meta-Analysis," *Journal of Mental and Nervous Disease* 202 (2014): 353–59.

57. Rodney Dietert, "Integrating Contemplative Tools into Biomedical Science Education and Research Training Programs," *Journal of Biomedical Education* 2014 (2014): 1–11.

CHAPTER 8. BEYOND SCIENCE AND RELIGION

1. Margaret Mead, *Coming of Age in Samoa: A Psychological Study of Primitive Youth for Western Civilisation* (New York: William Morrow, 1928), 1.

2. His Holiness the Dalai Lama, *Ethics for the New Millennium* (New York: Riverhead Books, 1999).

3. His Holiness the Dalai Lama, *Beyond Religion: Ethics for a Whole World* (New York: Houghton Mifflin Harcourt, 2011).

4. Paul Root Wolpe, Interview with PBS, *Religion and Ethics Newsweekly*, February 21, 2014, www.pbs.org/wnet/religionandethics/2014/02/21/february-21-2014-paul-root-wolpe-extended-interview/22244/.

INDEX

carbohydrates, 24, 83, 84
carbon dioxide, 152
cells, 11, 25, 42, 78; awareness in, 20, 32; as the basic unit of life, 19; B-cells, 205, 206; boundary membranes of, 20; brain cells, 122; cell activity and the connections of preexisting neurons, 63; cell signaling, 21–22; cytoplasm of, 120; dendritic memory cells, 205; division of, 59, 62; division of labor among, 32–33; fruiting body cells, 51; growth and division cycles of, 36–37 (*see also* mitosis); learning at the cellular level, 60–61; multicellular organisms, 30–31; overproduction of, 60; protein "sensors" within, 20–21; skin cells, 116; spore cells, 51; stalk cells, 51; T-cells, 205–6; unique environments of, 35–36; as a universe unto themselves, 11, 26; weapon cells, 58–59. *See also* cells, death of (apoptosis); consciousness, as an emergent property of cells; differentiation; human cells; multicellularity; neurons (nerve cells); stem cells
cells, death of (apoptosis), 51, 52, 56–57, 66, 67; occurrence of in the webbing between fingers and toes, 57; regulated cell death, 40
cerebral palsy, 203
children, social skills of, 186–87
clay(s), characteristics of, 77–78
Clinton, Bill, 110, 111
Cognitive-Based Compassion Training (CBCT), 191–92, 197; assessing the effects of, 215–19; as an example of a meditation method, 197–98, 208, 214; and ritualized breathing, 214;

teaching of to medical students and scientists, 219–20
communication, 28, 30; intercellular, 29; interorganism, 29; intracellular, 29
community, 45, 104, 157; importance of according to Dewey, 155; Tibetan community, 5, 164, 184, 191
consciousness: differing Buddhist concepts of, 169–70; as an emergent property of cells, 168–69
conservation, 28
Copernicus, 81
cortisol. *See* stress, and the cortisol stress response
creationism, 73, 99
"critical slowing," 148–49
Cure Within, The: A History of Mind-Body Medicine (Harrington), 192, 194
cycling, 28, 147

Dadul, Geshe, 24–25, 156–57, 191
Dalai Lama, 89, 102, 133, 157, 160, 178, 194, 196, 233; acceptance of evolution by, 74; on altruism, 229–30; on the collaboration of monks and scientists, 163–64; on compassion, 228; emissaries of to Tibetan communities in India, 190–91; escape of from Tibet, 16; on humans as social animals, 228; on the issue of sacrifice versus selfishness, 46; on the relationship between science and religion, 225, 226, 227; relationship of with Mao Zedong, 225–26; on risk, 73–74; on "secular ethics," 227–28, 229; on the uncovering of assumptions, 44; views on animal research, 15–16, 17
Darwin, Charles, 39, 70–71, 78, 81, 83,

87, 90–92, 171, 210; abhorrence of slavery, 82; concept of common ancestry, 89; delay in publishing due to fear of the church, 72

Darwin's Sacred Cause (Desmond and Moore), 82

Davidson, Richard, 185, 196

death, 40, 66; as part of life's stages, 142–43. *See also* cells, death of (apoptosis)

depression, 122, 125, 203, 211, 213–14, 217, 218; and the concept of panarchy, 146–49; major depression, 141, 145, 146–48, 203

Desmond, Adrian, 82

determination, 49

de Waal, Frans, 173–74

Dewey, John, 154; on the importance of community, 155

Dharamsala, 16, 41, 59, 191; computer lab of, 85; graduation ceremonies at, 221; two types of translation occurring at, 18

Dhondup, 33–34, 124, 128

Dias, Brian, 62, 63, 115

Dietert, Rodney, 220–21

differentiation, 48, 49

diversity, 28, 88, 113, 125, 136, 149; of bacteria, 203, 204, 205, 206; of birds, 90; of functions, 144; genetic diversity, 109; increase in, 54, 113

DNA, 22–23, 30, 54, 58, 63, 84–85, 116, 205; action of natural selection on, 88; alterations in (*see* mutations); DNA/protein sequences, 88; mitochondrial DNA, 52–53; mutation in bacterial DNA, 35; sequence of, 84, 118; sequencing technologies, 31, 111; single DNA base-pair change, 121; viral DNA mobilization, 113

Dobzhansky, Theodosius, 69, 74, 78; on the consonance of science and religion, 72–73

Dove, Rita, 190

E. coli, 204

ecology: and Buddhism, 130–31, 135, 141–42. *See also* Tibetan Plateau, ecosystem of

education: Bronfenbrenner's "ecological structure of the educational environment," 155; contemplative education, 158; as an ecosystem, 154–55; science education, 157–58; transformative education, 158

egg cells, in females, 40–41, 54; death of, 56; developing, 55; production of, 55–56

electricity, 160, 177, 179

elephants, and the practice of "targeted help," 176

embryogenesis, 57

emergent properties, 26, 27–29, 150–51, 163, 165, 167–68

Emory-Tibet Science Initiative, 141–42, 156, 170, 184, 191, 193, 199, 200, 219; and the Tenzin Gyatso Scholars, 189

emotions: evolution of, 171–74; negative emotions as necessary, 173

empathy, 181–82; and brain structure, 186; evolution of empathic behaviors, 177–78, 183–84; human capacity for, 183–84; impact of social context on, 182–83; integration of empathy and compassion, 188; key to the understanding of, 180–81; research concerning, 185

energy flow, 28

environment: the "big environments," 105, 108; collaboration of with genes, 88; developmental environment, 108–9; environment-organism interaction, 72; expanded definition of to include microbiomes, 206; gene-environment conversation, 104–5; microenvironments, 108; Western narrative of as separate from people, 140

enzymes, 22, 23, 24, 27

with viral DNA (viral sequences), 111–12; dynamism of, 110; and the encoding of RNA, 110; LINE-1 in genomes of individual cells, 113

geshe ceremonies, 158–59

giants, original human beings as, 85–86

Gide, André, 44

global warming, 152–53, 184

globalization, 184

glucocorticoid receptor gene expression, 117

glucose, 24

Gunderson, Lance, 141, 142, 145

Han Chinese, 83, 84, 106, 134, 140; birthing difficulty of at high altitudes, 80

Harrington, Anne, 192–93, 197; on the Eastward journeys, 194–96

Hart, Mickey, 191

hearing, 29

hippocampus, 61, 62, 119, 186, 205

Holling, C. S., 141; "adaptive management" concept of, 155–56

homeostasis, 19, 28, 202, 207

homosexuality, 180

Hooke, Robert, 11

HPA axis, 119–20, 122, 209

human beings: advantage of due to language, 91; all humans as members of the same species, 82; coevolution of with pathogens, 212–13; as electrical beings, 160; interrelatedness of with our ecosystem, 144–45; possible distinctions from other organisms, 79–81; relation to monkeys, 81, 87, 90; universal relatedness of, 81, 86–87

human cells, 14, 22, 30, 31, 35, 37, 52; hair cells, 37; liver cells, 28; skin cells, 37

hydrogen, 26, 28, 150

hypothalamus, 119, 205, 209, 218

immune system(s), 32, 57–58, 59, 209; effects of the lack of in mice, 205–6; innate immune system, 58; integral link of with the nervous system in humans, 206; life-and-death balance of immune-system cells, 60–61; and meditation, 215–18; and the "new" diseases, 199–203

impermanence, 103, 107, 131, 212; change as the nature of all impermanent phenomena, 144–45

incense, 125, 126

India, 178; emissaries of the Dalai Lama to Tibetan communities in, 190–91

infection, 58, 59, 139, 199, 206, 207, 209; cellular level memory of, 60–61; susceptibility to, 204; viral infection, 32, 216

inflammation: as a common result of disease, 209; decreasing inflammation, 209, 212; effect of Cognitive-Based Compassion Training (CBCT) on, 216–17; inflammation-causing molecules released by macrophages, 212; inflammation control, 208; "inflammatory bias," 213; inflammatory molecules, 205, 212, 214; and the inflammatory reflex, 208–10; role of the vagus nerve in, 210–12

"Inflammatory Reflex, The" (Tracey), 208

interdependence, 8, 28, 111, 130, 140, 149, 227; as a fundamental law of nature, 172

interleukin-6 (IL-6), 203, 205

slavery, 82

slime mold cells, 42–43, 46–48; response of to starvation, 50–52

social behaviors, 33

Spallanzani, Lazzaro, 74

species, evolution of new from old, 90–91

sperm: "activated," 40; death of, 40; dysfunctional, 53; production of in men, 53–54

squid, and glowing bacteria, 33

stem cells: division of, 67; research concerning, 126; role of in organ regeneration, 67–68; routine replenishment of, 67

stories: personal stories and the teaching of science, 124–28; as social support, 123–24

stress: and the cortisol stress response, 120–21, 209, 215; stress level, 121; "stress message," 120; and telomerase, 217–18; the "triangle of stress" experiments in monkeys, 202–4

structure/function relationship, 21, 28

sucrose, 24

sugar molecules, 21–22, 23–24

sunyata, 19–20

symbiotic relationships, 33

teaching, 28, 165–66; as an ecosystem model, 156–57; of evolution, 70, 72, 74, 76, 84–85, 96, 97–100, 156; personal stories and the teaching of science, 124–28; of science to Buddhist monks, 18, 34, 129, 140–41; and student input/concerns, 124–6; as a type of translation, 18–19, 157

telomerase, 217–18

Tenzin, story of, 101–2, 105–6, 108, 109; hypothetical identical twin of, 114, 122

Tibet, 178; government of in exile, 41–42; parallels of with the "settling" of the Western United States, 140

Tibetan Buddhism, 37, 184, 227; and ecology, 130–31; and modern science, 87–88, 231–32

Tibetan Plateau, ecosystem of, 131–33, 139–40, 141, 144; effect of global warming on, 152–53; farming and hunting in, 134–35; glaciers of, 134; high-altitude pastures and grasslands of, 136; pastoral life of the nomads of, 136, 139–40; predicting the tipping point of, 151. *See also* yaks

Tibetans, 84, 106, 213; challenges faced by, 153–54; community of, 5, 164, 184, 191; family tree of, 82–83; high-altitude adaptations of, 80; possible reasons for the resilience of, 200–201; trauma among, 103–4

timeframes: circadian, 34–35; developmental, 34, 35; evolutionary, 34, 35

tipping elements, 149

tipping-point model, 144, 148, 150, 151, 203

Tracey, Kevin, 208–9

translation: and classroom communication, 28; and teaching, 18–19, 23

Trier Social Stress Test, 215

turtles, shell origin theories of, 78–79

twins, identical (Ramu and Shamu), 114, 122

vagus nerve, 210–12; communication of to the immune system through macrophages, 212; three levels of across evolutionary time, 211

Vinaya, 164